From Plant Data
to Process Control

From Plant Data to Process Control

Ideas for process identification and PID design

Liuping Wang

and

William R. Cluett

London and New York

First published 2000
by Taylor & Francis
11 New Fetter Lane, London EC4P 4EE

Simultaneously published in the USA and Canada
by Taylor & Francis Inc,
29 West 35th Street, New York, NY 10001-2299

Taylor & Francis is an imprint of the Taylor & Francis Group

© 2000 Liuping Wang and William R. Cluett

Printed and bound in Great Britain by T J International,
Padstow, Cornwall

Publisher's Note
This book has been prepared from camera-ready-copy supplied by the authors.

British Library Cataloguing in Publication Data
A catalogue record for this book is available from the British Library.

Library of Congress Cataloging in Publication Data

ISBN 0-7484-0701-4

To Jianshe and Robin (LW)

To Janet, Shannon, Taylor, Owen
    and my Mom and Dad (WRC)

Contents

Series Introduction

Control systems has a long and distinguished tradition stretching back to nineteenth-century dynamics and stability theory. Its establishment as a major engineering discipline in the 1950s arose, essentially, from Second World War-driven work on frequency response methods by, amongst others, Nyquist, Bode and Wiener. The intervening 40 years has seen quite unparalleled developments in the underlying theory with applications ranging from the ubiquitous PID controller, widely encountered in the process industries, through to high-performance fidelity controllers typical of aerospace applications. This development has been increasingly underpinned by the rapid developments in the, essentially enabling, technology of computing software and hardware.

This view of mathematically model-based systems and control as a mature discipline masks relatively new and rapid developments in the general area of robust control. Here an intense research effort is being directed to the development of high-performance controllers which (at least) are robust to specified classes of plant uncertainty. One measure of this effort is the fact that, after a relatively short period of work, 'near world' tests of classes of robust controllers have been undertaken in the aerospace industry. Again, this work is supported by computing hardware and software developments, such as the toolboxes available within numerous commercially marketed controller design/simulation packages.

Recently, there has been increasing interest in the use of so-called 'intelligent' control techniques such as fuzzy logic and neural networks. Basically, these rely on learning (in a prescribed manner) the input–output behaviour of the plant to be controlled. Already, it is clear that there is little to be gained by applying these techniques to cases where mature mathematical model-based approaches yield high-performance control. Instead, their role (in general terms) almost certainly lies in areas where the processes encountered are ill-defined, complex, nonlinear, time-varying and stochastic. A detailed evaluation of their (relative) potential awaits the appearance of a rigorous supporting base (underlying theory and implementation architectures, for example) the essential elements of which are beginning to appear in learned journals and conferences.

Elements of control and systems theory/engineering are increasingly finding use outside traditional numerical processing environments. One such general area is intelligent command and control systems which are central, for example, to innovative manufacturing and the management of advanced transportation systems. Another is discrete event systems which mix numeric and logic decision making.

It was in response to these exciting new developments that the book series on *Systems and Control* was conceived. It publishes high-quality research texts and reference works in the diverse areas which systems and control now includes. In addition to basic theory, experimental and/or application studies are welcome, as are expository texts where theory, verification and applications come together to provide a unifying coverage of a particular topic or topics.

E. Rogers
J. O'Reilly

Acknowledgements

This book brings into one place the work we carried out together over the period 1989–1998 in the field of process identification and control. Much of the book focuses on two model structures for dynamics systems, the Laguerre model and the Frequency Sampling Filter (FSF) model. We were introduced to the Laguerre model by Zervos and Dumont (1988). We first encountered the FSF model in Middleton (1988). With respect to the Laguerre model, our interest arose quite naturally from the fact that Guy Dumont was one of the principal investigators in the Government of Canada's Mechanical Wood Pulps Network of Centres of Excellence which provided us with the funding to begin our collaboration. In the case of the FSF model, we just feel lucky to have picked up the Proceedings of the IFAC Workshop on Robust Adaptive Control in which Rick Middleton's paper appeared. These two papers look at using the Laguerre and FSF models in an adaptive control context so perhaps this was a factor as well, given that both of our doctoral dissertations were in this field. Although this book contains nothing specifically dealing with adaptive control, our interest in identification and control obviously started there.

This work was carried out with the help of many excellent students. We would like to acknowledge the contributions made by Nirmala Arifin, Tom Barnes, Michelle Desarmo, Errol Goberdhansingh, Xunqing Jiang, Alex Kalafatis, Marshall Khan, Althea Leitao, Sophie McQueen, Sharon Pate, Umesh Patel, Dianne Smektala, Joe Tseng and Alex Zivkovic. We also acknowledge the generous funding provided by the Natural Sciences and Engineering Research Council of Canada through the Wood Pulps Network and a Collaborative Research and Development Grant, along with our industrial partners Imperial Oil Ltd, Sunoco Group and Noranda Inc. Finally, we would like to thank Stephen Woo and Leonard Segall (Imperial Oil), Cliff Pedersen and Mike Foley (Sunoco), Roger Jones (Noranda), Bill Bialkowski (EnTech) and Alex Penlidis (University of Waterloo) for their encouragement and support.

Chapter 1

Introduction

1.1 THE LAGUERRE MODEL: PROCESS IDENTIFICATION FROM STEP RESPONSE DATA

An identification experiment consists of perturbing the process input and observing the resulting response in the process output variable. A process model describing this dynamic input-output relationship can then be *identified* directly from the data iself. In a process control context, the end-use of such a model would typically be for controller design.

The step response test is one of the simplest identification experiments to perform. The test involves increasing or decreasing the input variable from one operating point to another in a step fashion, and recording the behaviour of the output variable. Step response tests are often performed in industry in order to determine approximate values for the process gain, time constant and time delay (Ljung, 1987). However, these experiments are widely viewed as only a precursor to the design of further experiments, the collection of more input-output data, and the subsequent analysis of this data using regression-based techniques to obtain a more accurate model. However, the simplicity of the original step response test provides the incentive to fully explore the extent to which an accurate model may be obtained directly from the step response data itself.

Various methods are available in the literature for obtaining a transfer function model directly from step response data. For example, in Rake (1980) and Unbehauen and Rao (1987), graphical methods based on flexion tangents or times to reach certain percentage values of the final steady state are presented. An implicit requirement of these methods is that the step response data be relatively noise-free to the extent that the engineer can clearly *see* the true process response to the step input change. However, this is not the case in many practical situations.

Given the limitations of the graphical methods, we have chosen to approach this problem from a different perspective. Our objective is to devise a systematic algorithm that works directly with the step response data to produce a continuous-time, transfer function model of the process with minimum error in a least squares sense. We are able to achieve this goal in Chapter 2 by taking advantage of the orthonormal properties of the Laguerre functions, which have received considerable attention in the recent literature on system identification and automatic control (Zervos and Dumont, 1988; Mäkilä, 1990; Wahlberg, 1991; Wahlberg and Ljung, 1992; Goodwin *et al.*, 1992).

The proposed method for estimating the parameters of this Laguerre model is simple and straightforward, involving only numerical integration of the step response data. One of the most important features of the Laguerre model is its time scaling factor, p. If this parameter is selected suitably, the Laguerre model can be used to efficiently approximate a large class of linear systems. Clowes (1965) illustrated how to select the optimal time-scaling factor for systems with rational transfer functions, assuming that an analytic expression for the system's impulse response is available and that there is no delay present in the process. We extend Clowes' result to a general class of stable linear systems and propose a simple strategy for determining the optimal time scaling factor directly from the step response data. An analysis of the effect of disturbances occurring during the step response test on the model quality is also presented. We classify various types of disturbances based on their frequency content, and identify the types which have a significant impact on the quality of the estimated model. We also perform this analysis in the time domain and use this to show that a simple pretreatment of the step response data can greatly enhance the accuracy of the estimated model.

The above analysis shows that, as long as the disturbances are *fast* relative to the process dynamics, an accurate model can in fact be constructed from step response data. However, many processes are affected by slow, drifting disturbances that effectively *mask* the true process response. For these types of disturbances, the proposed Laguerre approach may produce process models with significant errors. In this case, other types of input signals, such as a random binary input signal or a periodic input signal, should be used to enable the effect of the disturbances on the process output to be separated from the process response due to the input variable.

1.2 USE OF PRESS FOR MODEL STRUCTURE SELECTION IN PROCESS IDENTIFICATION

When working with regression-based techniques for process model identification, one of the challenging tasks is to determine the most appropriate process model structure. In a linear model context, this would be information such as the number of poles and zeros to be included in the transfer function description. If the structure of the system being identified is known in advance, then the problem reduces to a much simpler parameter estimation problem.

Cross-validation is often recommended in the literature as a technique for determining the most appropriate model structure (Ljung, 1987; Korenberg *et al.*, 1988). With cross-validation, the data set generated from the identification experiment is split into an estimation set, which is used to estimate the parameters, and a testing set, which is used to judge the predictive capability of the model. This step is particularly useful in revealing the structure of a dynamic system subject to disturbances where it is believed that the disturbance sequence will never be exactly duplicated from the estimation set to the testing set.

There is another way to generate the prediction errors without actually having to split the data set. The idea is to set aside each data point, estimate a model using the rest of the data, and then evaluate the prediction error at the point that was removed. This concept is well known as the *PRESS* statistic in the statistical community (Myers, 1990) and is used as a technique for model validation of general regression models. However, to our knowledge, the system identification literature has not suggested the use of the *PRESS* for model structure selection.

Chapter 3 presents the development of the *PRESS* statistic as a criterion for structure selection of dynamic process models which are linear-in-the-parameters. Computation of the *PRESS* statistic is based on the orthogonal decomposition algorithm proposed by Korenberg *et al.* (1988) and can be viewed as a by-product of their algorithm since very little additional computation is required. We also show how the *PRESS* statistic can be used as an efficient technique for noise model development directly from time series data.

1.3 FREQUENCY SAMPLING FILTERS: AN IMPROVED MODEL STRUCTURE FOR PROCESS IDENTIFICATION

For the industrial application of multivariable model predictive process control, the dynamic relationships between the manipulated inputs and controlled outputs are typically expressed in terms of high order finite impulse response (FIR) or finite step response (FSR) models relating each input to each output. These models fall in a class which we will refer to as *input-only* models where the process output is expressed as a function of only past values of the process input. The FIR/FSR models are popular because they fit very naturally into the predictive control algorithms and also because the types of multivariable processes on which these controllers are typically applied are not well represented by lower order transfer function models (Cutler and Yocum, 1991; MacGregor *et al.*, 1991). The FIR/FSR models are also appealing because they are a straightforward representation of the process dynamics.

Despite these advantages, there are a few widely recognized problems associated with the identification of these FIR/FSR models from process input-output data. The first problem is their high dimensionality. The order of these models is equal to the settling time of the process (the time required for the process output to reach a new steady state after a change has been made in the process input) divided by the data sampling interval. Therefore, FIR/FSR model orders of at least 50 to 100 are not unusual. The second problem is that these model structures often result in ill-conditioned solutions when applying a least squares estimator. The optimal input signal for identifying an FIR model is one containing rich excitation at all frequencies (Levin, 1960). However, this kind of input signal is seldom used in the process industries. The types of test signal more often used consist of relatively infrequent input moves. As a result, the data matrices associated with the estimation of the FIR models are often poorly conditioned which inflates the variance of the parameter estimates and, as a result, leads to nonsmooth FIR models.

To overcome these problems, MacGregor *et al.* (1991) have looked at biased regression techniques (e.g. ridge regression (RR)) and the projection to latent structures (PLS) method as alternatives to least squares. Ricker (1988) studied the use of PLS and a method based on the singular value decomposition (SVD). All of these approaches attempt to reduce the parameter variances and improve the numerical stability of the solution with the tradeoff being biased models.

Recognizing that the reason for lack of smoothness of the FIR models lies with the type of input signals used for identification experiments in the process industries, we have chosen to focus on an alternative model structure for process identification in Chapters 4 and 5. Our approach is fundamentally different from the RR, PLS and SVD approaches in the sense that we approach this problem by first performing a frequency decomposition of the identified model, separating low and medium frequency parameters from high frequency parameters and then by choosing to ignore these high frequency parameters in the final model structure. This frequency decomposition is based on the frequency sampling filter (FSF) model, which is simply a linear transformation of the FIR model. Therefore, it maintains the main advantage of the FIR model in that it requires no structural information about the process, such as its order and relative degree. The FSF structure was first introduced to the areas of system identification and automatic control by Bitmead and Anderson (1981), Parker and Bitmead (1987) and Middleton (1988).

In the new FSF model parameter estimation problem, the delayed values of the process input that appear in the data matrices for estimating the FIR model are replaced by filtered values of the process input, where the filters have very narrow band-limited characteristics. Also, the discrete process impulse response weights, which represent the parameters of the FIR model, are replaced by the discrete process frequency response coefficients. These narrow band-limited filtered input signals separate the frequency components of the input signal and yield a least squares correlation matrix that has diagonal elements proportional to the power spectrum of the input. When the input spectrum has little content in the frequency range of estimation, the correlation matrix becomes ill-conditioned. Therefore, the problem of smoothing the step response estimates is converted into identifying the optimal number of frequency sampling filters to be included in the FSF model. This optimal number can be found by examining the model's predictive capability, e.g. as measured by the *PRESS* statistic presented in Chapter 3. Alternatively, because the number of FSF model parameters needed to accurately represent many process step responses is often far fewer than the number required by an FIR model, and because this number is independent of the sampling interval, we have also found that we can safely fix the number of frequency sampling filters and hence the number of FSF model parameters to be estimated at a modest level, say 11 or 13, for a large class of systems.

1.4 PID CONTROLLER DESIGN: A NEW FREQUENCY DOMAIN APPROACH

The PID controller continues to be the most common type of single-loop feedback regulator used in the process industries. However, the tuning of these controllers is still not widely understood and, in fact, many still operate with their original default settings. Despite this, researchers continue to strive to find relatively simple ways to design these controllers in order to improve closed-loop performance. However, it is safe to say that not one method in over 50 years has been able to replace the Ziegler-Nichols (1942) tuning methods in terms of familiarity and ease of use.

More recent developments in the area of PID controller tuning fall into three categories:

Model-Based Designs

A structured model of the process (typically a Laplace transfer function) is used directly in a design method such as pole-placement or internal model control (IMC) to yield expressions for the controller parameters that are functions of the process model parameters and some user-specified para-meter related to the desired performance, e.g. a desired closed-loop time constant. These approaches to PID design carry restrictions on the allow-able model structure, although it has been shown that a wide range of types of processes can be accommodated if the PID controller is augmented with a first order filter in series. An example of this design approach may be found in Rivera *et al.* (1986).

Designs Based on Optimization of an Integral Feedback Error Performance Criterion

This approach can be applied to a wide variety of transfer function models. However, a numerical search procedure is required to find the optimal con-troller parameters. See, for example Zhuang and Atherton (1993).

Designs Based on Process Frequency Response

Perhaps motivated by the popular Ziegler-Nichols frequency response method which requires knowledge of only one point on the process Nyquist curve, ways have been developed to automate the Ziegler-Nichols method (Åström and Hägglund, 1984), to refine their tuning formulae (Hang *et al.*, 1991) and to develop improved design methods which require only a slight in-crease in the amount of process frequency response information (Åström and Hägglund, 1988; Åström, 1991).

From our point of view, each approach has its advantages. The first two model-based approaches have a more intuitive time domain performance specification than traditional frequency domain design methods. However, the frequency domain methods require less structural information about the process dynamics. Chapters 6 and 7 present a new frequency domain PID design approach that we feel combines these advantages. This new design method begins with a time domain performance specification on the behaviour of the closed-loop *control signal* rather than a specification on the desired output signal or feedback error. The behaviour of the controller output is an important consideration when assessing overall closed-loop performance in a process control application (Harris and Tyreus, 1987). In addition, we propose to use only a limited number of points on the process Nyquist curve for controller design without requiring any structural information about the process dynamics other than knowledge of whether or not the process is self-regulating. Since we make use of points on the process Nyquist curve in the design, we address the question of which frequency response points have the largest impact on the closed-loop time domain performance and therefore which should be used in the design. Here, we exploit the connection between the frequency domain and the time domain made in our earlier work with the FSF model in Chapters 4 and 5. Straightforward analytical solutions for the PID parameters, or tuning rules, are also derived for first order plus delay and integrating plus delay processes in order to put our results on a comparable footing with other PID tuning formulae in terms of ease of use. These tuning rules contain a single closed-loop response speed parameter to be selected by the user.

1.5   RELAY FEEDBACK EXPERIMENTS FOR PROCESS IDENTIFICATION

The relay feedback experiment was made popular in the field of process control by Åström and Hägglund (1984). This experiment was suggested as a means to automate the Ziegler-Nichols scheme for determining ultimate gain and frequency information about a process. Their approach followed directly from a describing function approximation (DFA) to the nonlinear relay element. The objective was to use the obtained process information for automatic tuning of PID controllers.

Åström and Hägglund's work (1984) has prompted research in several different directions. One of these directions, and the focus of Chapter 8, is in the area of process identification, where the objective is to obtain a more complete and accurate model of the process from data generated under relay feedback. Fitting a more complete process model (i.e. a transfer function model) normally requires knowledge of several points on the process Nyquist curve. Given that the standard relay experiment combined with the DFA identification technique is able to identify only a single point, fitting such a model either requires the availability of some prior process information (e.g. Luyben, 1987) or requires the user to conduct a series of relay experiments in which the oscillation frequency is adjusted by incorporating various dynamic elements into the relay feedback loop (e.g. Li *et al.*, 1991; Schei, 1994).

In Chapter 8, it is shown that the frequency sampling filter (FSF) model along with a least squares estimator can be used in conjunction with the data generated from a standard relay experiment to quickly and accurately identify the process frequency response at the dominant harmonics of the limit cycle. A recursive implementation of the least squares algorithm is suggested for parameter estimation. This methodology is extended by introducing a modified relay experiment designed to enable the identification of a more complete process step response model from a single relay experiment. In this experiment, the error signal is switched back and forth between a standard relay element and an integrator in series with a relay. The generated input signal is no longer periodic as in the case of the standard relay experiment, but instead is typically rich in the frequency range needed for accurate step response model identification. Because this method makes use of the FSF model structure, the only required prior process knowledge is an estimate of the process settling time and it will be demonstrated that even this information may be estimated directly from the modified relay experiment.

Chapter 2

Modelling from Noisy Step Response Data Using Laguerre Functions

2.1 INTRODUCTION

This chapter introduces a method for building Laplace transfer function models from noisy step response data. The algorithm is based on the Laguerre functions and exploits their orthonormal properties to produce a simple, yet effective approach.

This chapter contains seven sections plus an appendix. Section 2.2 presents the Laguerre functions, describes how they may be used to develop a transfer function model of a process (called the Laguerre model), and defines the Laguerre coefficients in both the time domain and frequency domain. Section 2.3 refines a classic optimization approach for selecting the time scaling factor in the Laguerre model. Section 2.4 introduces the step response modelling algorithm, in which the model coefficients and the optimal time scaling factor are estimated directly from the step response data itself. Section 2.5 analyzes the statistical properties of the estimated coefficients, leading to the conclusion that their variances are related to the power spectrum of the disturbance. Section 2.6 further analyzes the errors associated with the estimated coefficients in the time domain and proposes a simple data pretreatment procedure that can be applied to the step response data to improve the model accuracy. In Section 2.7, the modelling algorithm is applied to step response data obtained from a pilot-scale polymerization reactor.

Portions of this chapter have been reprinted from *Chemical Engineering Science* **50**, L. Wang and W.R. Cluett, "Building transfer function models from noisy step response data using the Laguerre network", pp. 149-161, 1995, with permission from Elsevier Science, and from *IEEE Transactions on Automatic Control* **39**, L. Wang and W.R. Cluett, "Optimal choice of time-scaling factor for linear system approximations using Laguerre models", pp. 1463-1467, 1994, with permission from IEEE.

2.2   PROCESS REPRESENTATION USING LAGUERRE MODELS

This section introduces the Laguerre model for representing the process transfer function. The basic idea is to approximate the continuous-time impulse response of the process in terms of the orthonormal Laguerre functions. The Laguerre coefficients themselves will then be defined in terms of both the process impulse response and its frequency response.

2.2.1   Approximation of the process impulse response

A sequence of real functions $l_1(t), l_2(t), \ldots$ is said to form an orthonormal set over the interval $(0, \infty)$ if they have the property that

$$\int_0^\infty l_i^2(t)dt = 1 \tag{2.1}$$

and

$$\int_0^\infty l_i(t)l_j(t)dt = 0 \quad i \neq j \tag{2.2}$$

A set of orthonormal functions $l_i(t)$ is called complete if there exists no function $f(t)$ with $\int_0^\infty f(t)^2 dt < \infty$, except the identically zero function, such that

$$\int_0^\infty f(t)l_i(t)dt = 0 \tag{2.3}$$

for $i = 1, 2, \ldots$.

The Laguerre functions (Lee, 1960) are an example of a set of complete orthonormal functions that satisfy the properties defined by Equations (2.1)-(2.3). The set of Laguerre functions is defined as, for any $p > 0$

$$l_1(t) = \sqrt{2p} \times e^{-pt}$$
$$l_2(t) = \sqrt{2p}(-2pt + 1)\, e^{-pt}$$

$$l_3(t) = \sqrt{2p}(+2p^2t^2 - 4pt + 1) \; e^{-pt}$$

$$l_4(t) = \sqrt{2p}(-\frac{4}{3}p^3t^3 + 6p^2t^2 - 6pt + 1) \; e^{-pt} \qquad (2.4)$$

$$l_5(t) = \sqrt{2p}(+\frac{2}{3}p^4t^4 - \frac{16}{3}p^3t^3 + 12p^2t^2 - 8pt + 1) \; e^{-pt}$$

$$\vdots = \vdots$$

$$l_i(t) = \sqrt{2p}[(-1)^{i-1}\frac{(2p)^{i-1}}{(i-1)!}t^{i-1} + (-1)^i\frac{(i-1)(2p)^{i-2}}{(i-2)!}t^{i-2}$$

$$+(-1)^{i-1}\frac{(i-1)(i-2)(2p)^{i-3}}{2!(i-3)!}t^{i-3} + \cdots + 1]e^{-pt}$$

The parameter p is called the time scaling factor for the Laguerre functions. This parameter plays an important role in their practical application and will be discussed in detail in Section 2.3. (Note: The set of Laguerre functions presented in Equations (2.4) differs by a factor of -1 for even values of i when compared with the set of Laguerre functions presented by Lee (1960). However, this does not affect the orthonormal properties of these functions.)

Definition of Coefficients in the Time Domain

With respect to a set of functions $l_i(t)$ that is orthonormal and complete over the interval $(0, \infty)$, it is known that an arbitrary function $h(t)$ has a formal expansion analogous to a Fourier expansion (Wylie, 1960). Such an expansion has been widely used in numerical analysis for the approximation of functions in differential and integral equations. The idea behind using Laguerre functions to represent a linear, time invariant process is to take $h(t)$ to be the unit impulse response of the process, where $h(t)$ can be written as

$$h(t) = c_1 l_1(t) + c_2 l_2(t) + \cdots + c_i l_i(t) + \cdots \qquad (2.5)$$

and $\{c_i\}$ are the coefficients of the expansion defined by

$$c_1 = \int_0^\infty l_1(t)h(t)dt$$

$$c_2 = \int_0^\infty l_2(t)h(t)dt \qquad (2.6)$$

$$\vdots = \vdots$$

$$c_i = \int_0^\infty l_i(t)h(t)dt$$

Convergence Condition in Time Domain

The expansion given in Equation (2.5), in theory, requires an infinite numbers of terms in order for it to converge to the true impulse response. How-

ever, the assumed completeness of the set of orthonormal functions ensures
that, for any piecewise continuous impulse response function $h(t)$ with

$$\int_0^\infty h^2(t)dt < \infty \tag{2.7}$$

and any $\varepsilon > 0$, which is a measure of the accuracy of the approximation,
there exists an integer N such that the integral squared error between the
true and approximated impulse responses is less than ε, i.e.

$$\int_0^\infty (h(t) - \sum_{i=1}^N c_i l_i(t))^2 dt < \varepsilon \tag{2.8}$$

Therefore, we can use a truncated expansion $\sum_{i=1}^N c_i l_i(t)$ to closely approximate the unit impulse response $h(t)$ with an increasing number of terms,
N.

2.2.2   Approximation of the process transfer function

In parallel with the above time domain description, an approximation of the
process transfer function using the Laguerre functions can also be developed.
The Laplace transform of the impulse response $h(t)$ in Equation (2.5) leads
to the continuous-time transfer function of the process

$$\begin{aligned}
G(s) &= \int_0^\infty h(t)e^{-st}dt \\
&= \int_0^\infty (c_1 l_1(t) + c_2 l_2(t) + \cdots + c_i l_i(t) + \cdots)e^{-st}dt \\
&= c_1 L_1(s) + c_2 L_2(s) + \cdots + c_i L_i(s) + \cdots \tag{2.9}
\end{aligned}$$

where the Laplace transforms of the Laguerre functions, also referred to as
the Laguerre filters, are given by

$$\begin{aligned}
L_1(s) &= \int_0^\infty l_1(t)e^{-st}dt = \frac{\sqrt{2p}}{(s+p)} \\
L_2(s) &= \int_0^\infty l_2(t)e^{-st}dt = \frac{\sqrt{2p}(s-p)}{(s+p)^2} \\
\vdots \; &= \; \vdots \\
L_i(s) &= \int_0^\infty l_i(t)e^{-st}dt = \frac{\sqrt{2p}(s-p)^{i-1}}{(s+p)^i}
\end{aligned} \tag{2.10}$$

The process transfer function given by Equations (2.9) and (2.10) is called
the Laguerre model. The Laguerre filters in Equation (2.10) have a simple

form that is easy to remember in that the filters have all their poles at the same location, $-p$, and all their zeros at $+p$. The first filter, $L_1(s)$, is a first order low-pass filter. All other filters, $L_i(s)$, consist of a first order filter, $L_1(s)$, in series with an all-pass filter $[\frac{s-p}{s+p}]^{i-1}$. (Note: The Laguerre filters presented in Equations (2.10) differ from the Laguerre filters presented by Lee (1960) in that the numerator of the general ith filter in Lee (1960) is $(p-s)^{i-1}$ instead of $(s-p)^{i-1}$. However, our presentation is consistent with that used by Zervos and Dumont (1988).)

Definition of Coefficients in the Frequency Domain

Parseval's theorem (Desoer and Vidyasagar, 1975) states that, if two real functions $x(\tau)$ and $y(\tau)$ are bounded in the l_2 space (namely $\int_{-\infty}^{\infty} x(\tau)^2 d\tau < \infty$ and $\int_{-\infty}^{\infty} y(\tau)^2 d\tau < \infty$), then

$$\int_{-\infty}^{\infty} x(\tau)y(\tau)d\tau = \frac{1}{2\pi} \int_{-\infty}^{\infty} X^*(jw)Y(jw)dw \qquad (2.11)$$

where $X(jw)$ and $Y(jw)$ are the Fourier transforms of $x(\tau)$ and $y(\tau)$, and f^* denotes the complex conjugate of f. Application of Parseval's theorem to Equations (2.1) and (2.2) gives the orthonormal properties in the frequency domain

$$\frac{1}{2\pi} \int_{-\infty}^{\infty} |L_i(jw)|^2 dw = 1 \qquad (2.12)$$

and

$$\frac{1}{2\pi} \int_{-\infty}^{\infty} L_i(jw)L_j^*(jw)dw = 0 \quad i \neq j \qquad (2.13)$$

The coefficients $\{c_i\}$ defined by Equations (2.6) can be expressed as

$$c_1 = \frac{1}{2\pi} \int_{-\infty}^{\infty} G^*(jw)L_1(jw)dw$$

$$c_2 = \frac{1}{2\pi} \int_{-\infty}^{\infty} G^*(jw)L_2(jw)dw \qquad (2.14)$$

$$\vdots = \vdots$$

$$c_i = \frac{1}{2\pi} \int_{-\infty}^{\infty} G^*(jw)L_i(jw)dw$$

Convergence Condition in the Frequency Domain

Condition (2.7), that permitted the use of a truncated expansion to closely approximate the unit impulse response $h(t)$, may also be given in the frequency domain by direct application of Parseval's theorem as

$$\frac{1}{2\pi} \int_{-\infty}^{\infty} |G(jw)|^2 dw < \infty \qquad (2.15)$$

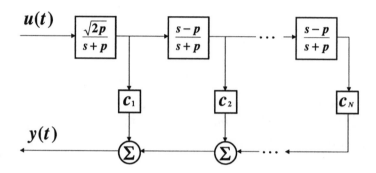

Figure 2.1: *Laguerre network*

which implies that the process transfer function $G(s)$ has all its poles strictly on the left half of the complex plane and has a strictly proper structure (i.e. $\lim_{w\to\infty} |G(jw)| = 0$). The latter condition holds when the order of the transfer function numerator is less than the order of the denominator. Processes that satisfy this condition are referred to as L_2 stable systems.

2.2.3   Laguerre model in state space form

Figure 2.1 shows the block diagram of the Laguerre model (order N) described by Equations (2.9) and (2.10), where $u(t)$ is the process input and $y(t)$ is the process output. The process input passes through the Laguerre filters arranged in series and the filter outputs are weighted by their respective Laguerre coefficients. The sum of these weighted filtered signals gives the process output $y(t)$.

From this block diagram, we can derive the Laguerre model in its state space form. Defining the state vector

$$X(t) = [l_1(t) \quad l_2(t) \quad \cdots \quad l_N(t)]^T$$

and assuming zero initial conditions of the state vector, then

$$
\begin{bmatrix} \dot{l}_1(t) \\ \dot{l}_2(t) \\ \vdots \\ \dot{l}_N(t) \end{bmatrix} = \begin{bmatrix} -p & 0 & \cdots & 0 \\ -2p & -p & \cdots & 0 \\ \vdots & & & \vdots \\ -2p & \cdots & -2p & -p \end{bmatrix} \begin{bmatrix} l_1(t) \\ l_2(t) \\ \vdots \\ l_N(t) \end{bmatrix} + \sqrt{2p} \begin{bmatrix} 1 \\ 1 \\ \vdots \\ 1 \end{bmatrix} u(t) \quad (2.16)
$$

$$
y(t) = [c_1 \quad c_2 \quad \cdots \quad c_N] X(t) \quad (2.17)
$$

2.2.4 Generating the Laguerre functions

It is important to be able to efficiently generate values for the Laguerre functions. There are several ways to do so and each way requires a different amount of computational effort.

Method A. For low model orders, the Laguerre functions can be generated using Equations (2.4) directly.

Method B. For higher model orders, a recursive approach proposed in Atkinson (1989) can be used.

Method C. When MATLAB is available, the transfer functions of the Laguerre filters can be used to evaluate their unit impulse responses.

Method D. The set of differential equations in Equation (2.16) can be solved numerically.

We will now give more detailed information about Method B and Method D.

Method B: Generating Laguerre Functions Using Polynomials

Atkinson (1989) presents the following recursive relation between what are known as the Laguerre polynomials denoted here by P_i, where

$$
P_0(x) = 1
$$

$$
P_1(x) = 1 - x
$$

and, for any index number $n \geq 1$,

$$
P_{n+1}(x) = \frac{1}{n+1}(2n + 1 - x)P_n(x) - \frac{n}{n+1}P_{n-1}(x)
$$

We can now generate the Laguerre functions in Equations (2.4) by setting $x = 2pt$ in the Laguerre polynomials

$$
\begin{aligned}
l_1(t) &= P_0(2pt)\sqrt{2p}e^{-pt} \\
l_2(t) &= P_1(2pt)\sqrt{2p}e^{-pt} \\
\vdots &= \vdots \\
l_i(t) &= P_{i-1}(2pt)\sqrt{2p}e^{-pt}
\end{aligned}
\tag{2.18}
$$

Method D: Generating Laguerre Functions Using Difference Equations

By examining Equation (2.16), we find that the Laguerre functions in Equations (2.4) satisfy the following set of differential equations

$$
\begin{bmatrix} \dot{l}_1(t) \\ \dot{l}_2(t) \\ \vdots \\ \dot{l}_N(t) \end{bmatrix} =
\begin{bmatrix} -p & 0 & \cdots & 0 \\ -2p & -p & \cdots & 0 \\ & \vdots & & \vdots \\ -2p & \cdots & -2p & -p \end{bmatrix}
\begin{bmatrix} l_1(t) \\ l_2(t) \\ \vdots \\ l_N(t) \end{bmatrix}
\tag{2.19}
$$

with the initial conditions

$$
\begin{bmatrix} l_1(0) \\ l_2(0) \\ \vdots \\ l_N(0) \end{bmatrix} = \sqrt{2p}
\begin{bmatrix} 1 \\ 1 \\ \vdots \\ 1 \end{bmatrix}
\tag{2.20}
$$

Hence, the solution of this set of differential equations yields the time domain Laguerre functions, which can be found numerically by iteratively solving the following set of difference equations

$$
\begin{bmatrix} l_1(t_{i+1}) \\ l_2(t_{i+1}) \\ \vdots \\ l_N(t_{i+1}) \end{bmatrix} =
\begin{bmatrix} -p & 0 & \cdots & 0 \\ -2p & -p & \cdots & 0 \\ & \vdots & & \vdots \\ -2p & \cdots & -2p & -p \end{bmatrix}
\begin{bmatrix} l_1(t_i) \\ l_2(t_i) \\ \vdots \\ l_N(t_i) \end{bmatrix} \times \Delta t +
\begin{bmatrix} l_1(t_i) \\ l_2(t_i) \\ \vdots \\ l_N(t_i) \end{bmatrix}
\tag{2.21}
$$

with

$$
\begin{bmatrix} l_1(t_0) \\ l_2(t_0) \\ \vdots \\ l_N(t_0) \end{bmatrix} = \sqrt{2p}
\begin{bmatrix} 1 \\ 1 \\ \vdots \\ 1 \end{bmatrix}
\tag{2.22}
$$

and $\Delta t = t_{i+1} - t_i$ being the integration step size. As long as Δt is sufficiently small, this numerical scheme is stable and produces sufficiently accurate solutions.

2.3   CHOICE OF THE TIME SCALING FACTOR

In theory, choice of the time scaling factor p does not affect the existence and convergence of the Laguerre model with respect to the model order N. The accuracy of the approximation increases with increasing model order. In practice though, a poor choice of p requires a high order Laguerre model in order to achieve a desired model accuracy. However, the estimation of an accurate Laguerre model from process data corrupted by noise and disturbances becomes more difficult when using a large value for N. Therefore, one of the keys to the successful application of the Laguerre modelling approach is to find a systematic method for optimizing the choice of the time scaling factor p. To demonstrate the importance of this issue, an illustrative example is given.

Example 2.1. Consider the construction of a Laguerre model for the first order process described by

$$G(s) = \frac{a}{s+a} \tag{2.23}$$

where $a > 0$. The unit impulse response of this process is given by

$$h(t) = ae^{-at} \tag{2.24}$$

We can compute the coefficients of the Laguerre model using Equations (2.14) for a positive time scaling factor p

$$
\begin{aligned}
c_1 &= \frac{1}{2\pi} \int_{-\infty}^{\infty} \frac{a}{-jw+a} \sqrt{2p} \frac{1}{jw+p} dw \\
&= \frac{a\sqrt{2p}}{a+p}
\end{aligned}
\tag{2.25}
$$

$$
\begin{aligned}
c_2 &= \frac{1}{2\pi} \int_{-\infty}^{\infty} \frac{a}{-jw+a} \sqrt{2p} \frac{(jw-p)}{(jw+p)^2} dw \\
&= \frac{a\sqrt{2p}}{a+p} \frac{a-p}{a+p}
\end{aligned}
\tag{2.26}
$$

and

$$c_i = \frac{1}{2\pi} \int_{-\infty}^{\infty} \frac{a}{-jw+a} \sqrt{2p} \frac{(jw-p)^{i-1}}{(jw+p)^i} dw$$

$$= \frac{a\sqrt{2p}}{a+p} \frac{(a-p)^{i-1}}{(a+p)^{i-1}} \qquad (2.27)$$

Therefore, the *Nth* order Laguerre model for this first order system is

$$\hat{G}(s) = \frac{a(2p)}{a+p} [\frac{1}{s+p} + \cdots + \frac{(a-p)^{N-1}}{(a+p)^{N-1}} \frac{(s-p)^{N-1}}{(s+p)^N}] \qquad (2.28)$$

We can see from Equation (2.28) that this Laguerre model serves only as an approximation to the original system unless $p = a$.

2.3.1   Modelling errors with respect to choice of p

The integral squared error between the unit impulse response of the process and that of the *Nth* Laguerre model is defined as

$$E = \int_0^{\infty} (h(t) - \sum_{i=1}^{N} c_i l_i(t))^2 dt \qquad (2.29)$$

and the derivative of this integral squared error with respect to c_i is given by

$$\frac{dE}{dc_i} = -2 \int_0^{\infty} [h(t) - \sum_{i=1}^{N} c_i l_i(t)] l_i(t) dt \qquad (2.30)$$

Using the orthonormal properties of the Laguerre functions, Equation (2.30) is equivalent to

$$\frac{dE}{dc_i} = -2 \int_0^{\infty} [h(t) l_i(t) - c_i l_i(t) l_i(t)] dt \qquad (2.31)$$

and by setting $\frac{dE}{dc_i} = 0$, we find that

$$c_i = \int_0^{\infty} l_i(t) h(t) dt \qquad (2.32)$$

Equation (2.32) corresponds to the original definition of the Laguerre coefficients in Equations (2.6). It can be shown that the solutions of the coefficients given by Equation (2.32) minimize the integral squared error in

Equation (2.29) because the second derivative of E with respect to c_i is always positive.

The integral squared error in Equation (2.29) can be rewritten as

$$
\begin{aligned}
E &= \int_0^\infty h(t)^2 dt - 2 \int_0^\infty \sum_{i=1}^N c_i l_i(t) h(t) dt + \int_0^\infty \left(\sum_{i=1}^N c_i l_i(t) \right)^2 dt \\
&= \int_0^\infty h(t)^2 dt - \sum_{i=1}^N c_i^2 \qquad\qquad (2.33)
\end{aligned}
$$

where the expressions for the Laguerre coefficients given by Equation (2.32) and the orthonormal properties of the Laguerre functions have been used.

Using Parseval's theorem, Equation (2.29) can be also represented in the frequency domain as

$$
E = \frac{1}{2\pi} \int_{-\infty}^\infty \left| G(jw) - \sum_{i=1}^N c_i L_i(jw) \right|^2 dw \qquad\qquad (2.34)
$$

The expressions for the Laguerre coefficients in Equations (2.14) in terms of the process frequency response can be derived by minimizing Equation (2.34). The error E can also be expressed in a form similar to Equation (2.33), where

$$
E = \frac{1}{2\pi} \int_{-\infty}^\infty |G(jw)|^2 dw - \sum_{i=1}^N c_i^2 \qquad\qquad (2.35)
$$

The first term on the right-hand side of Equation (2.33) or Equation (2.35) is independent of the time scaling factor p and therefore only the second term is a function of p. Hence, for a given model order N, the minimum error E with respect to p corresponds to the maximum of $\sum_{i=1}^N c_i^2$ with respect to p. Therefore, the problem of searching for an optimal time scaling factor p is converted to finding the maximum of the loss function defined by

$$
V = \sum_{i=1}^N c_i^2 \qquad\qquad (2.36)
$$

2.3.2 Optimal choice of p

The optimal choice of the time scaling factor p described by Clowes (1965) is generalized here for any L_2 stable system.

Theorem 2.1: Given that the Laguerre coefficients $\{c_i\}$ can be obtained from Equations (2.14), and assuming that the true system $G(s)$ is L_2 stable, then the derivative of the loss function V with respect to the time scaling factor p is given by

$$\frac{dV}{dp} = \frac{d(\sum_{i=1}^{N} c_i^2)}{dp} = \frac{N}{p} c_N c_{N+1} \tag{2.37}$$

To prove the theorem, we first require the following lemma.

Lemma 2.1: For some $p > 0$, the Laplace transforms of the Laguerre functions given in Equations (2.10) satisfy the following equality

$$2p \frac{d(L_i(s))}{dp} = i L_{i+1}(s) - (i-1) L_{i-1}(s) \tag{2.38}$$

Proof of Lemma 2.1: It can be readily shown that for $i = 1$

$$\frac{d(L_1(s))}{dp} = \frac{1}{\sqrt{2p}} \frac{s-p}{(s+p)^2} = \frac{1}{2p} L_2(s) \tag{2.39}$$

and for $i = 2$

$$2p \frac{d(L_2(s))}{dp} = 2 L_3(s) - L_1(s) \tag{2.40}$$

Now assume that for $i > 3$, the following equality is true

$$2p \frac{d(L_{i-1}(s))}{dp} = (i-1) L_i(s) - (i-2) L_{i-2}(s) \tag{2.41}$$

Therefore, we must demonstrate that

$$2p \frac{d(L_i(s))}{dp} = i L_{i+1}(s) - (i-1) L_{i-1}(s) \tag{2.42}$$

Using Equations (2.10)

$$L_i(s) = L_{i-1}(s) \frac{s-p}{s+p} \tag{2.43}$$

we can write

$$2p \frac{d(L_i(s))}{dp} = 2p \left[\frac{d(L_{i-1}(s))}{dp} \frac{s-p}{s+p} - L_{i-1}(s) \frac{2s}{(s+p)^2} \right] \tag{2.44}$$

and substituting Equation (2.41) into Equation (2.44) leads to

$$
\begin{aligned}
2p\frac{d(L_i(s))}{dp} &= [(i-1)L_i(s) - (i-2)L_{i-2}(s)]\frac{s-p}{s+p} - L_{i-1}(s)\frac{4ps}{(s+p)^2} \\
&= iL_{i+1}(s) - (i-2)L_{i-1}(s) - L_{i-1}(s)\left[\frac{4ps}{(s+p)^2} + \frac{(s-p)^2}{(s+p)^2}\right] \\
&= iL_{i+1}(s) - (i-1)L_{i-1}(s)
\end{aligned}
\tag{2.45}
$$

which proves the lemma by induction.

Proof of Theorem 2.1: Note that from Equations (2.14)

$$
2p\frac{dc_i}{dp} = \frac{2p}{2\pi}\int_{-\infty}^{\infty} G^*(jw)\frac{dL_i(jw)}{dp}\,dw
\tag{2.46}
$$

Applying Lemma 2.1 gives

$$
2p\frac{dc_i}{dp} = ic_{i+1} - (i-1)c_{i-1}
\tag{2.47}
$$

which is equivalent to

$$
2c_i\frac{dc_i}{dp} = \frac{1}{p}[ic_ic_{i+1} - (i-1)c_ic_{i-1}]
\tag{2.48}
$$

Considering that

$$
\frac{d(\sum_{i=1}^{N} c_i^2)}{dp} = 2\sum_{i=1}^{N} c_i\frac{dc_i}{dp}
\tag{2.49}
$$

then applying the summation to both sides of Equation (2.48) gives

$$
\begin{aligned}
\frac{d(\sum_{i=1}^{N} c_i^2)}{dp} &= \frac{1}{p}\sum_{i=1}^{N}[ic_ic_{i+1} - (i-1)c_ic_{i-1}] \\
&= \frac{N}{p}c_Nc_{N+1}
\end{aligned}
\tag{2.50}
$$

which proves the theorem.

Remarks:

- The problem of finding a maximum of $\sum_{i=1}^{N} c_i^2$ with respect to p reduces to finding the zeros of either of the coefficients c_N or c_{N+1} as a function of p and then checking that the value of c_Nc_{N+1} changes sign from positive to negative as p increases. Each value of p corresponding to a maximum can then be used to evaluate the actual value of $\sum_{i=1}^{N} c_i^2$ in order to determine the optimal time scaling factor p.

- For a given model order N, Equation (2.37) tells us that we are looking for the value of p that corresponds to a zero of c_{N+1}. Otherwise, the model order could be reduced to $N-1$ without any change in model accuracy (i.e. $c_N = 0$).

- From Equation (2.47), it can be verified that

$$\frac{d^2 c_i}{dp^2} = \frac{1}{4p^2}[i(i+1)c_{i+2} - 2ic_{i+1} - (i^2 + (i-1)^2)c_i$$
$$+ 2(i-1)c_{i-1} + (i-1)(i-2)c_{i-2}] \tag{2.51}$$

Both the first and second derivatives in Equations (2.47) and (2.51) are useful for applying numerical methods to find the zeros of the coefficients, once an interval is located in which a zero is known to exist.

Example 2.2. Irrational transfer functions have been approximated in the literature using truncated infinite partial fraction expansions (Partington *et al.*, 1988) and the Lagrange interpolation formula (Olivier, 1992). Here, we will illustrate that this class of linear systems can be efficiently approximated by a Laguerre model based on the minimization of the frequency domain loss function in Equation (2.34). We will consider the following system (Partington *et al.*, 1988)

$$G(s) = \frac{1}{s + 1 - e^{-s-2}} \tag{2.52}$$

Our objective is to approximate this system using a $3rd$ order Laguerre model ($N = 3$). In order to find the optimal time scaling factor p, we have computed the coefficients c_3 and c_4 based on Equations (2.14) for a range of p values and have noted that the product $c_3 c_4$ only changes sign from positive to negative in the interval $(0.9, 1.2)$. Hence, the optimal value of p is located in this region. Applying Newton's method, we found the optimal p to be equal to 1.09 with a corresponding value of $c_4 = -9.2849 \times 10^{-6}$. The first three Laguerre coefficients are given as

$$c_1 = 7.2159 \times 10^{-1}$$

$$c_2 = -8.3489 \times 10^{-2}$$

$$c_3 = 4.7121 \times 10^{-2}$$

leading to the following Laguerre model

$$\hat{G}(s) = \frac{1.0117s^2 + 2.1709s + 1.4949}{(s + 1.09)^3} \tag{2.53}$$

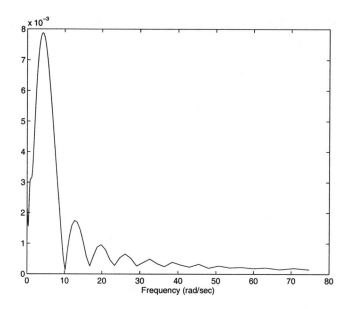

Figure 2.2: *Magnitude of frequency domain error for Example 2.2*

The quality of this model can be measured by its L_∞ norm error

$$\max_{w_i} |G(jw_i) - \hat{G}(jw_i)| = 7.8799 \times 10^{-3} \qquad (2.54)$$

or by its L_2 norm error

$$\sum_{i=1}^{\infty} |G(jw_i) - \hat{G}(jw_i)|^2 = 2.6373 \times 10^{-3} \qquad (2.55)$$

where $10^{-7} \le w_i \le 10^5$ and we have used 500 logarithmically equally spaced frequencies in this region to evaluate these errors. Figure 2.2 shows the magnitude of the frequency domain error. It is interesting to note that although the Laguerre model is obtained by minimizing an L_2 norm error, the resulting L_∞ norm error is actually slightly smaller than the L_∞ norm error of 7.9×10^{-3} associated with the $23rd$ order partial fraction expansion model given by Partington *et al.* (1988).

The choice of the time scaling factor p is crucial in this example in terms of its effect on both the L_∞ norm error and L_2 norm error. For example, for $p = 0.9$, $\max_{w_i} |G(jw_i) - \hat{G}(jw_i)| = 1.2924 \times 10^{-2}$ and $\sum_{i=1}^{\infty} |G(jw_i) - \hat{G}(jw_i)|^2 = 4.559 \times 10^{-2}$, and for $p = 1.2$, $\max_{w_i} |G(jw_i) - \hat{G}(jw_i)| = 1.2679 \times 10^{-2}$ and $\sum_{i=1}^{\infty} |G(jw_i) - \hat{G}(jw_i)|^2 = 4.5826 \times 10^{-2}$.

2.3.3   Optimal time scaling factor for first order plus delay systems

First order plus delay systems are commonly encountered in the process industries and therefore it is important to consider the choice of an optimal time scaling factor for this class of systems. Our intention is to derive some empirical rules based on the process time delay and time constant so that a near optimal time scaling factor can be found with little computational effort.

Example 2.1 illustrated that the optimal value of p for a first order system is equal to the inverse of the process time constant. If the process is higher order but without time delay, satisfactory results can be obtained if p is chosen based on the dominant time constant of the process. However, the presence of delay can greatly affect the optimal choice of p. To examine this problem, we shall first derive an analytical solution for the Laguerre coefficients associated with a first order plus delay system and then find empirical rules for choosing the optimal time scaling factor p.

Laguerre Coefficients

The transfer function of a first order plus delay system is given by

$$G(s) = \frac{Ka}{s+a}e^{-ds} \tag{2.56}$$

where K is the process gain, $\frac{1}{a}$ is the process time constant and d is the process delay. The impulse response of the process is given by

$$h(t) = Ke^{-a(t-d)} \tag{2.57}$$

for $t \geq d$, and $h(t) = 0$ for $0 \leq t < d$. In this case, it is convenient to evaluate the Laguerre coefficients in the time domain. Using the Laguerre functions in matrix form by defining $X(t) = [l_1(t) \quad l_2(t) \quad \dots \quad l_N(t)]^T$ and the $N \times N$ matrix

$$A = \begin{bmatrix} 1 & 0 & \dots & 0 \\ 2 & 1 & \dots & 0 \\ \vdots & & & \\ 2 & \dots & 2 & 1 \end{bmatrix} \tag{2.58}$$

the solution of Equation (2.19) leads to

$$X(t) = exp(-pAt)X(0) \tag{2.59}$$

with $X(0) = \sqrt{2p}[1 \quad 1 \quad \cdots \quad 1]^T$. Thus, the Laguerre coefficient vector, $C = [c_1 \quad c_2 \quad \cdots \quad c_N]^T$, is given by

$$C = Ke^{ad} \int_d^\infty e^{-at} X(t)dt \qquad (2.60)$$

Using integration by parts

$$\int_d^\infty e^{-at} X(t)dt = \frac{1}{a}e^{-ad}X(d) - \frac{p}{a}A \int_d^\infty e^{-at} X(t)dt \qquad (2.61)$$

Therefore the analytical solution of the Laguerre coefficients for the first order plus delay process is

$$C = K(aI + pA)^{-1}X(d) \qquad (2.62)$$

Derivation of Empirical Rules

The first step toward derivation of empirical rules for the optimal choice of the time scaling factor p is to reduce the number of variables in Equation (2.62) from 3 (a, p and d) to 2. To do so, we will choose to let $\gamma = ad$ and $p_d = pd$. Then, combining Equations (2.59) and (2.62) gives

$$C = K\sqrt{2p} \times d(\gamma I + p_d A)^{-1}e^{-Ap_d} \begin{bmatrix} 1 \\ 1 \\ \vdots \\ 1 \end{bmatrix} \qquad (2.63)$$

We are interested in finding the zeros of the coefficient vector C with respect to p_d for different values of γ. The special structure of the A matrix in Equation (2.58) allows us to write down polynomial expressions in terms of p_d for the coefficients, and from these expressions, to directly solve for the zeros of the coefficients. To determine the optimal p value, the roots corresponding to negative and complex values of p_d are discarded and the optimal p is identified by examining the behaviour of $c_N c_{N+1}$. One additional point to note from Equation (2.63) is that the gain of the first order plus delay process does not affect the optimal pole location.

We have attempted to develop some empirical algebraic expressions for the optimal choice of p by examining the behaviour of $c_N c_{N+1}$ up to $N = 3$. We have studied systems with $0 < \gamma \le 1.5$, e.g. a system with $\gamma = 1.5$ has a delay that is 1.5 times larger than its time constant. In the region $0 < \gamma \le 0.303$, the optimal time scaling parameter is determined by one of the zeros of the third coefficient, while for $0.303 \le \gamma \le 1.5$, it is determined by one of the zeros of the fourth coefficient. We have found that for both regions of γ, the optimal time scaling factor p only changes slightly as the model order is increased beyond $N = 3$. Therefore, this value of p will be near its optimal value, regardless of the final order of the model. We have numerically obtained the optimal values of p for the two regions of γ identified above and have then fit polynomials as a function of γ to yield the following empirical solutions

$$p \;=\; \frac{1}{d}(-4.1151\gamma^2 + 3.3670\gamma - 0.0572) \quad 0.045 \leq \gamma < 0.303 \quad (2.64)$$

$$p \;=\; \frac{1}{d}(0.7546\gamma^3 - 2.7505\gamma^2 + 3.7643\gamma + 0.1452)$$
$$0.35 \leq \gamma \leq 1.5 \tag{2.65}$$

For γ values greater than 1.5, we have found that it is better to select the optimal time scaling factor by examining the zeros of Equation (2.63) for the chosen value of N.

Example 2.3. Consider the following 12*th* order process

$$G(s) = \frac{(15s + 1)^2(4s + 1)(2s + 1)}{(20s + 1)^3(10s + 1)^3(5s + 1)^3(0.5s + 1)^3} \tag{2.66}$$

This system can be approximated by a first order plus delay model structure with a delay $d = 20$ and $a = 1/50$ directly from its step response shown in Figure 2.3. Thus, with $\gamma = 0.4$ in this case, an estimate of the optimal time scaling factor p is obtained from Equation (2.65) as $p = 0.063$. The coefficients of the Laguerre model can be computed directly from the process frequency response $G(jw)$ using Equations (2.14). Choosing $N = 6$ gives the following Laguerre model

$$\hat{G}(s) = \frac{B(s)}{A(s)} \tag{2.67}$$

where $A(s) = (s + 0.063)^6$ and $B(s) = -5.6197 \times 10^{-4}s^5 + 2.8905 \times 10^{-4}s^4 - 6.5453 \times 10^{-5}s^3 + 5.6715 \times 10^{-6}s^2 + 1.5646 \times 10^{-6}s + 6.2345 \times 10^{-8}$. The step response of this Laguerre model is compared with the step response of the true system in Figure 2.3 and the two are indistinguishable. The magnitude of the frequency domain error between the true system and the Laguerre model is shown in Figure 2.4. The maximum error is given by $\max_w |G(jw) - \hat{G}(jw)| = 2.88 \times 10^{-3}$. It is worth noting that the optimal time scaling factor for this example with $N = 4$, obtained by examining the

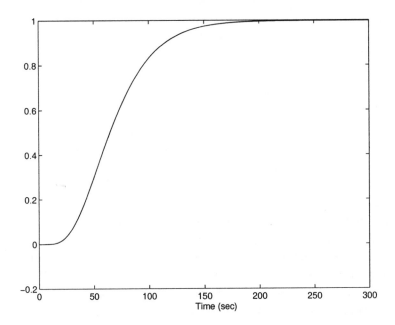

Figure 2.3: *Step response of the 12th order system for Example 2.3*

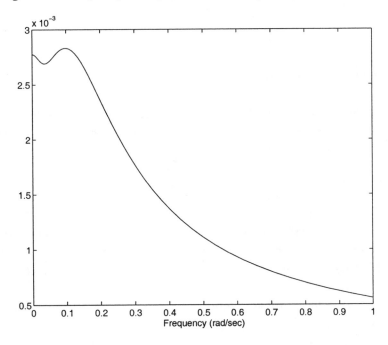

Figure 2.4: *Magnitude of frequency domain error for Example 2.3*

zeros of c_4c_5, is equal to 0.0625. Beyond $N = 4$, we found that the optimal value for p did not change significantly and therefore the estimate obtained from the empirical solution is very close to the optimal one.

2.4   ESTIMATION OF LAGUERRE COEFFICIENTS FROM STEP RESPONSE DATA

Looking at Equations (2.6), impulse response data might appear to be an obvious choice for estimating the coefficients of the Laguerre model. An impulse response test required to generate such a data set involves the application of an impulse input signal to the process, which in theory is a signal with infinite amplitude for zero time duration. In practice, this type of signal can only be approximately realized using a pulse signal with a very large amplitude and short duration. However, a large amplitude input change may not be realizable or acceptable. On the other hand, a small amplitude input move may result in a very low signal to noise ratio. For these reasons, the impulse response test is seldom performed in practice.

In contrast, the step response test is frequently performed in the process industries. A step response test uses a step input signal defined by

$$
\begin{aligned}
u(t) \quad &= \quad 0 \quad t < 0 \\
&= \quad u_m \quad t \geq 0
\end{aligned}
$$

where the magnitude of the step change u_m is chosen by the user. This test is simple and easy to perform.

The relationship between the measured process step response $\hat{g}(t)$ and the corresponding estimated process impulse response $\hat{h}(t)$ is given by

$$
\hat{h}(t) = \frac{d\hat{g}(t)}{dt}
\tag{2.68}
$$

Although we could use the estimate for $\hat{h}(t)$ presented in Equation (2.68) in place of $h(t)$ in Equations (2.6), differentiation of the measured step response will amplify the noise effects and cause numerical problems. It would be preferable to avoid differentiation and work directly with the measured step response instead.

Assume that for a unit step input change, the output response of a stable, linear, time invariant system is given by

$$
\hat{g}(t) = g(t) + \xi(t)
\tag{2.69}
$$

where $g(t)$ denotes the true process step response (in deviation form), and $\xi(t)$ denotes the output additive disturbance with $|\xi(t)| < \infty$.

In a noisy environment, the process step response will not settle at a fixed steady state value. However, the mean of the new steady state value, g_{mean}, can be calculated using $g_{mean} = \frac{1}{T_{end}-T_s} \int_{T_s}^{T_{end}} \hat{g}(t)dt$, where T_s is an estimate of the time it takes the process to reach steady state, and T_{end} is the end time of the step response experiment.

Replacing $h(t)$ by $\frac{d\hat{g}(t)}{dt}$ in Equations (2.6) gives the estimate for the ith Laguerre coefficient

$$\hat{c}_i = \int_0^\infty \frac{d\hat{g}(t)}{dt} l_i(t)dt$$

$$= \int_0^\infty l_i(t)d\hat{g}(t) \tag{2.70}$$

Applying integration by parts, Equation (2.70) becomes

$$\hat{c}_i = [\hat{g}(t)l_i(t)]_0^\infty - \int_0^\infty \hat{g}(t)\dot{l}_i(t)dt \tag{2.71}$$

Since for any $p > 0$, $\lim_{t\to\infty} l_i(t) = 0$, and taking the initial value of the measured step response as $\hat{g}(0) = 0$, the first term on the right-hand side of Equation (2.71) is equal to zero. Thus

$$\hat{c}_i = -\int_0^\infty \hat{g}(t)\dot{l}_i(t)dt \tag{2.72}$$

This is the key equation for estimating the Laguerre coefficients from step response data, and it will be used subsequently for analysis of variance and bias of the estimates as well as for the development of a data pretreatment strategy later in this chapter. However, Equation (2.72) is still not in the final form to be used for computational purposes. Using the estimate of the process settling time T_s, Equation (2.72) can be rearranged into

$$\hat{c}_i = -\int_0^{T_s} \hat{g}(t)\dot{l}_i(t)dt - \int_{T_s}^\infty \hat{g}(t)\dot{l}_i(t)dt \tag{2.73}$$

Replacing $\hat{g}(t)$ in the interval (T_s, ∞) with g_{mean}, then

$$-\int_{T_s}^\infty \hat{g}(t)\dot{l}_i(t)dt = g_{mean}l_i(T_s) \tag{2.74}$$

The derivative of the ith Laguerre function from Equation (2.19) satisfies

$$\dot{l}_i(t) = -2pl_1(t) - 2pl_2(t) - \cdots - pl_i(t) \tag{2.75}$$

Substituting Equations (2.74) and (2.75) into Equation (2.73) gives the estimate of the *ith* coefficient as

$$\hat{c}_i = 2p \int_0^{T_s} \hat{g}(t)l_1(t)dt + 2p \int_0^{T_s} \hat{g}(t)l_2(t)dt + \cdots + p \int_0^{T_s} \hat{g}(t)l_i(t)dt + g_{mean}l_i(T_s)$$
(2.76)

The equations for estimating the Laguerre model coefficients using step response data can now be summarized as follows

$$
\begin{aligned}
\hat{c}_1 &= p \int_0^{T_s} \hat{g}(t)l_1(t)dt + g_{mean}l_1(T_s) \\
\hat{c}_2 &= 2p \int_0^{T_s} \hat{g}(t)l_1(t)dt + p \int_0^{T_s} \hat{g}(t)l_2(t)dt + g_{mean}l_2(T_s) \\
&\vdots \\
\hat{c}_i &= 2p \int_0^{T_s} \hat{g}(t)l_1(t)dt + 2p \int_0^{T_s} \hat{g}(t)l_2(t)dt + \cdots \\
&\quad + p \int_0^{T_s} \hat{g}(t)l_i(t)dt + g_{mean}l_i(T_s)
\end{aligned}
$$
(2.77)

Remarks:

- During the derivation of the estimation algorithm, integration by parts is used so that the derivative is transferred from the step response $\hat{g}(t)$ to the Laguerre functions. Here, the Laguerre function is being used as a type of modulating function in that the approximation of a derivative from a noisy signal is avoided (Unbehauen and Rao, 1987; Co and Ydstie, 1990).

- This algorithm does not involve the inversion of an input-output data matrix that usually arises with the application of least squares based estimation algorithms. Therefore, any numerical problems associated with inversion of this matrix when using a step input signal are avoided.

- Each estimated coefficient can be estimated independently from the values of the other estimated coefficients. Therefore, an increase or decrease in the model order N will not affect the previously estimated coefficients, for the same value of p. This differs from most other algorithms, where a change in model order would affect the values of all estimated parameters.

Estimation of Laguerre Coefficients from Discrete-Time Data

The estimation algorithm given by Equations (2.77) consists of the solution of a set of integral equations. In a modern data acquisition environment, the measured output step response will be a set of discrete-time data. Therefore, the solutions must be implemented using numerical integration schemes.

Let $\hat{g}(t_0), \hat{g}(t_1), \dots$ denote the discretized step response data with sampling interval $\Delta t = t_{i+1} - t_i$ for all i, and $t_0 = 0$. One approach for computing the Laguerre coefficients could be based on the rectangular rule. For instance, the integral equations found in Equations (2.77) would be approximated by

$$\int_0^{T_s} \hat{g}(t)l_i(t)dt \approx \Delta t \sum_{j=0}^{M-1} \hat{g}(t_j)l_i(t_j) \tag{2.78}$$

where $M = \frac{T_s}{\Delta t}$. Alternatively, Simpson's rule could be used to obtain more accurate estimates of the integral expressions, where

$$\int_0^{T_s} \hat{g}(t)l_i(t)dt \approx \frac{\Delta t}{3}[\hat{g}(t_0)l_i(t_0) + 4\hat{g}(t_1)l_i(t_1) + 2\hat{g}(t_2)l_i(t_2) + \cdots$$
$$+ 2\hat{g}(t_{2M-2})l_i(t_{2M-2}) + 4\hat{g}(t_{2M-1})l_i(t_{2M-1})$$
$$+ \hat{g}(t_{2M})l_i(t_{2M})] \tag{2.79}$$

with $M = \frac{T_s}{2\Delta t}$.

Choice of p Revisited

The first step in choosing the optimal time scaling factor p is to identify the interval in which it might be located. Using the estimate of the process settling time T_s, we can form an interval $[p_{min}, p_{max}]$, where the lower end of the interval, $p_{min} = \frac{4 \text{ to } 5}{T_s}$, is chosen to approximately correspond to the optimal value of p for a first order process with settling time T_s. The upper end of the interval, p_{max}, is chosen to be either $5p_{min}$ for a lower order Laguerre model fit ($N \leq 4$) or $10p_{min}$ for a higher order Laguerre model fit. We have found that such an interval generally covers the region in which the optimal p lies for most processes. Then, this interval is divided into a set of discrete values for the time scaling factor.

Using the process step response data and for a given model order N, the Laguerre coefficients $\hat{c}_N$ and $\hat{c}_{N+1}$ are estimated for each value of the time scaling factor in the interval. Then, the function $\hat{c}_N\hat{c}_{N+1}$ is evaluated at each value of p in the interval and the regions where $\hat{c}_N\hat{c}_{N+1}$ changes sign from positive to negative are identified. For each region, a linear interpolation is used to find the value of the time scaling factor p that makes $\hat{c}_N\hat{c}_{N+1} = 0$.

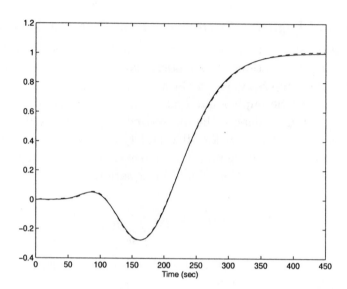

Figure 2.5: *Comparison of step responses for Example 2.4 (solid: true response; dashed: Laguerre model)*

That is, suppose $\hat{c}_N(p_i)\hat{c}_{N+1}(p_i) > 0$ and $\hat{c}_N(p_{i+1})\hat{c}_{N+1}(p_{i+1}) < 0$ for some p_i and p_{i+1}, then an approximate value of the time scaling factor p' in the interval $[p_i, p_{i+1}]$ where $\hat{c}_N(p')\hat{c}_{N+1}(p') = 0$ is equal to

$$p' = p_i - \hat{c}_N(p_i)\hat{c}_{N+1}(p_i)\frac{p_i - p_{i+1}}{\hat{c}_N(p_i)\hat{c}_{N+1}(p_i) - \hat{c}_N(p_{i+1})\hat{c}_{N+1}(p_{i+1})} \qquad (2.80)$$

Among all candidate values of p' found from Equation (2.80), we can identify the optimal p to be the one that produces the maximum value of $\sum_{i=1}^{N}\hat{c}_i^2$.

Example 2.4. Consider the process described by the transfer function

$$G(s) = \frac{(-45s + 1)^2(4s + 1)(2s + 1)}{(20s + 1)^3(18s + 1)^3(5s + 1)^3(10s + 1)^2(16s + 1)(14s + 1)(12s + 1)} \qquad (2.81)$$

This is a high order process with severe nonminimum phase behaviour. Its noise-free unit step response is shown in Figure 2.5. This process step response is sampled with an interval $\Delta t = 1.5$ sec. The key to success with the Laguerre model for such a complicated process is to find the optimal time scaling factor p for a given model order N, particularly when the model order is small. Figure 2.6 shows a 3-dimensional plot of the loss function $V = \sum_{i=1}^{N}\hat{c}_i^2$ for $N = 1, 2, \ldots, 10$ and $0 < p < 0.1$, where the coefficients

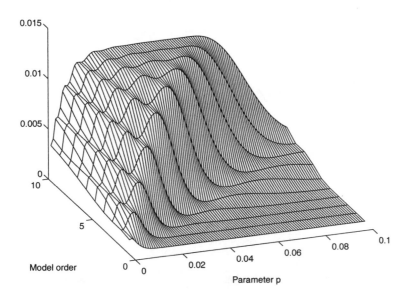

0.015
0.01
0.005
0
10
5
Model order
0
0
0.02
0.04
0.06
0.08
0.1
Parameter p

Figure 2.6: *3-dimensional plot of the loss function* $V = \sum_{i=1}^{N} \hat{c}_i^2$ *for Example 2.4*

$\hat{c}_i$, $i = 1, 2, \ldots, N$ are calculated using the algorithm given by Equations (2.77). This plot shows that the optimal time scaling factor p is found over a very narrow range when this complicated system is being described by a low order model. It also shows that as the model order N increases, the maximum of the loss function increases, i.e. the accuracy of the Laguerre model approximation improves. This increasing behaviour of the loss function gradually levels off once the model order reaches 8. Also, as the model order increases, the range of choices for an acceptable p value widens.

Suppose that we choose $N = 8$ to approximate this 14*th* order system. The optimal time scaling factor p can be found by examining the behaviour of $\hat{c}_8\hat{c}_9$ over the chosen interval $0 < p < 0.1$. Figure 2.7 shows this behaviour where it can be seen that $\hat{c}_8\hat{c}_9$ changes sign five times from positive to negative. After performing the necessary interpolations using Equation (2.80), the values of $\sum_{i=1}^{8} \hat{c}_i^2$ are calculated for the candidate values of p, and the optimal time scaling factor p is found to be 0.0501. Using this optimal time scaling factor, an 8*th* order Laguerre model is estimated to approximate the transfer function in Equation (2.81) using the step response data shown in Figure 2.5. This same figure shows the comparison between the true step response and the step response of the estimated Laguerre model. Figure 2.8 compares the frequency responses of the true process and the model.

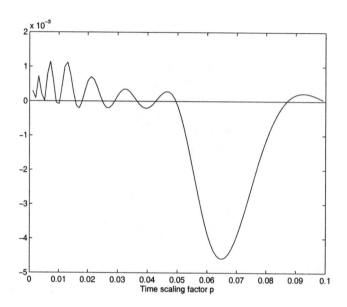

Figure 2.7: $\hat{c}_8\hat{c}_9$ as a function of p for Example 2.4

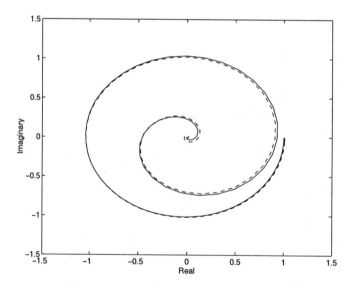

Figure 2.8: Comparison of frequency responses for Example 2.4 (solid: true response; dashed: Laguerre model)

2.5 STATISTICAL PROPERTIES OF THE ESTIMATED COEFFI-CIENTS

The step response modelling method presented in this chapter gives excellent results when the step response data is relatively noise-free, as illustrated in Example 2.4. However, the true challenge of any modelling method occurs when the process step response data is corrupted by significant disturbances. This section provides statistical analysis of the Laguerre model parameter estimates in both the frequency and time domains. Our analysis in the frequency domain shows that the accuracy of the parameter estimates is a function of the frequency content of the disturbance relative to the process dynamics. The time domain analysis leads naturally to a simple strategy for pretreating the step response data that improves the accuracy of the parameter estimates when disturbances are present.

2.5.1 Bias and variance analysis

Bias analysis: We assume that the measured unit step response can be represented by

$$\hat{g}(t) = g(t) + \xi(t)$$

where $g(t)$ is the noise-free step response and $\xi(t)$ is the additive disturbance. If the disturbance $\xi(t)$ has zero mean, then for any model order $N > 1$ and $p > 0$, the estimated coefficients $\hat{c}_i$, $i = 1, 2, \ldots, N$ are unbiased, i.e.

$$E[\hat{c}_i - c_i] = 0 \tag{2.82}$$

where $E[x]$ denotes the expected value of the variable x. Otherwise the estimated coefficients are biased.

Proof: The bias of the estimated Laguerre coefficient can be computed using Equations (2.72) and (2.75)

$$
\begin{aligned}
E[\hat{c}_i - c_i] &= 2p \int_0^\infty E[\hat{g}(t) - g(t)]l_1(t)dt + 2p \int_0^\infty E[\hat{g}(t) - g(t)]l_2(t)dt \\
&\quad + \cdots + p \int_0^\infty E[\hat{g}(t) - g(t)]l_i(t)dt \\
&= 2p \int_0^\infty E[\xi(t)]l_1(t)dt + 2p \int_0^\infty E[\xi(t)]l_2(t)dt \tag{2.83} \\
&\quad + \cdots + p \int_0^\infty E[\xi(t)]l_i(t)dt \tag{2.84}
\end{aligned}
$$

Then, unbiased results for the estimates follow from the condition that $E[\xi(t)] = 0$. It is also obvious that the estimated coefficients would be biased if $E[\xi(t)] \neq 0$.

Variance analysis: We assume that the disturbance $\xi(t)$ has zero mean with autocorrelation function $R_{\xi\xi}(\tau)$ and power spectrum $S_{\xi\xi}(w)$. We also define vectors $C = [c_1 \quad c_2 \quad \ldots \quad c_N]^T$ to contain the true Laguerre coefficients and $\hat{C}$ to contain the corresponding estimated coefficients. Then, the covariance of the estimated Laguerre coefficients is given by

$$E[(\hat{C} - C)(\hat{C} - C)^T] = p^2 A Q A^T \qquad (2.85)$$

where Q is a symmetric matrix with its elements defined by

$$Q_{ij} = \int_0^\infty l_j(t') \int_0^\infty l_i(s) R_{\xi\xi}(t' - s) ds dt' \qquad (2.86)$$

$$= \frac{1}{2\pi} \int_{-\infty}^\infty L_i(jw) S_{\xi\xi}(w) L_j^*(jw) dw \qquad (2.87)$$

and A is the $N \times N$ lower triangular matrix defined in Equation (2.58). In particular, the variance of the *ith* estimated coefficient is given by

$$Var[\hat{c}_i] = \frac{1}{\pi} \int_0^\infty W_i(\delta) S_{\xi\xi}(\delta) d\delta \qquad (2.88)$$

where

$$W_i(\delta) = 2p[1 - 2(-1)^{i-1} sin\theta \times sin[(2(i-1)+1)\theta] + sin^2\theta] \qquad (2.89)$$

with

$$\theta = tan^{-1}\frac{\delta}{p} \qquad (2.90)$$

Proof: see the Appendix in Section 2.8.

Remark:

- Equation (2.88) shows that, in the variance expression, $W_i(w)$ acts as a weighting function on the power spectrum of the disturbance $S_{\xi\xi}(w)$. We have plotted the weighting function $\frac{W_i(w)}{2p}$ for $i = 1, \ldots, 4$, in Figure 2.9. The parameter p acts as a scaling factor on both the frequency axis and the $W_i(w)$ axis. Thus, its value does not change the general shape of $W_i(w)$. However, as p is reduced, the amplitude and width of $W_i(w)$ decreases. As p is increased, $W_i(w)$ expands along both axes.

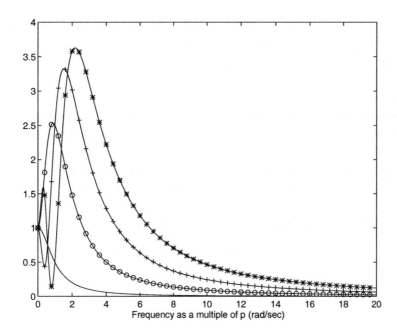

Figure 2.9: *Frequency domain weighting functions (solid: $i = 1$; solid with 'o': $i = 2$; solid with '+': $i = 3$; solid with '*': $i = 4$)*

In general, for a given disturbance spectrum $S_{\xi\xi}(w)$, the choice of a smaller value of p will reduce the variance of the estimated coefficients. The Laguerre coefficient index number i also affects both the shape and the maximum value of $W_i(w)$, i.e. the maximum value and the width of $W_i(w)$ increases as i increases. Hence, for a given disturbance spectrum, the variances of the estimated coefficients increase with the index i.

2.5.2 Some special cases of disturbances

By imposing some assumptions on the power spectrum of the disturbance, we can obtain explicit expressions for the variances of the estimated Laguerre coefficients.

Periodic Disturbances

Assume that $\xi(t) = D \cdot cos(w_0 t + \theta)$ with its corresponding power spectrum given by (Unbehauen and Rao, 1987)

$$S_{\xi\xi}(w) = \frac{D^2}{2}\pi\delta(w - w_0) \tag{2.91}$$

for $w \geq 0$, where δ is the Dirac delta function. Hence, from Equation (2.88), the variance of the *ith* estimated coefficient is

$$Var[\hat{c}_i] = \frac{D^2}{2} W_i(w_0) \tag{2.92}$$

which indicates that the variance is proportional to the squared amplitude of the periodic disturbance and to the value of the weighting function W_i at w_0. Since $\lim_{w \to \infty} W_i(w) = 0$, for all i, periodic disturbances at higher frequencies where $W_i(w)$ is small will produce small errors in the estimated coefficient $\hat{c}_i$ even if the disturbance amplitude D is large.

Disturbances with Band-Limited Spectrum

Assume that $\xi(t)$ is a disturbance with band-limited spectrum satisfying

$$
\begin{aligned}
S_{\xi\xi}(w) &= \int_{-\infty}^{\infty} R_{\xi\xi}(\tau) e^{-jw\tau} d\tau \\
&= k \quad w_1 < |w| < w_2 \tag{2.93} \\
&= 0 \quad otherwise \tag{2.94}
\end{aligned}
$$

Then the variance of the *ith* coefficient is given by

$$Var[\hat{c}_i] = \frac{k}{\pi} \int_{w_1}^{w_2} W_i(w) dw \tag{2.95}$$

Therefore, if the interval $[w_1, w_2]$ is in a higher frequency region where $W_i(w)$ is small, then the error in the estimated coefficient $\hat{c}_i$ will be small.

White Noise

If $\xi(t)$ is a continuous-time white noise sequence with autocorrelation function and power spectrum given by, for all τ and all w

$$R_{\xi\xi}(\tau) = k\delta(\tau); \; S_{\xi\xi}(w) = k \tag{2.96}$$

then the covariance of the estimated coefficients is given by

$$E[(\hat{C} - C)(\hat{C} - C)^T] = kp^2 A A^T \tag{2.97}$$

In particular, the variance of the estimate $\hat{c}_i$ is given by the *ith* diagonal element of Equation (2.97)

$$Var[\hat{c}_i] = k(4(i-1)+1))p^2 \tag{2.98}$$

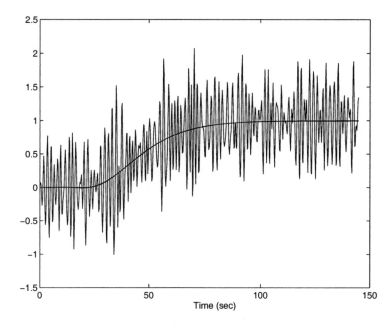

Figure 2.10: *Comparison of step responses for Example 2.5 (noisy signal: measured response; smooth signal: Laguerre model)*

Example 2.5. To illustrate the effect of disturbances on the accuracy of the Laguerre model coefficients, we examine the problem of estimating a Laguerre model from the following step response data

$$\hat{g}(t) = g(t) + \xi(t) \tag{2.99}$$

where $g(t)$ is generated from the unit step response of the following process

$$G(s) = \frac{e^{-20s}}{(10s + 1)^3} \tag{2.100}$$

and $\xi(t)$ is generated by passing a white noise sequence with unit variance through a band-pass filter in the frequency range $(0.4\pi, 0.8\pi)$ radians/sec. The process is sampled at an interval of 0.25 sec. The step response $\hat{g}(t)$ is illustrated in Figure 2.10.

The first step in estimating a Laguerre model for this process from this step response data is to determine the time scaling factor p and choose a model order N. It is seen from Figure 2.10 that the process has an approximate settling time of 100 sec, which gives us an estimate of the lower bound

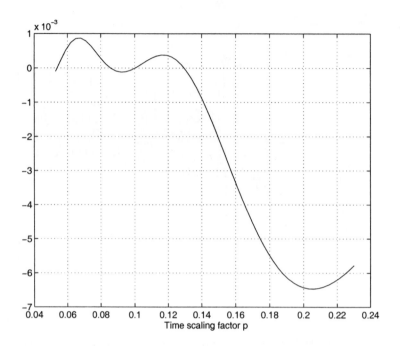

Figure 2.11: $\hat{c}_5\hat{c}_6$ *as a function of p for Example 2.5*

on the scaling factor as $\frac{5}{100} = 0.05$. The search interval is chosen to be $0.05 < p < 0.25$. Figure 2.11 shows the behaviour of $\hat{c}_5\hat{c}_6$. For a $5th$ order Laguerre model, the optimal time scaling factor is located at $p = 0.13$.

Table 2.1 lists the estimated coefficients of the $5th$ order Laguerre model obtained from the step response data shown in Figure 2.10, along with the coefficients obtained from the noise-free step response. The step response of the Laguerre model obtained from the noisy step response data is compared with the measured step response data in Figure 2.10 and the model's frequency response is compared in Figure 2.12 with the true process frequency response.

Since the disturbance power spectrum $S_{\xi\xi}(w)$ in this example is limited to a narrow frequency band and it can be shown that the spectrum does not significantly overlap any of the weighting functions $W_i(w)$ for $i = 1,\ldots,5$, it is expected from the variance analysis for the band-limited noise case (Equation (2.95)) that the accuracy of the estimated coefficients should not not be significantly compromised by the disturbance. This is confirmed by the results given in Table 2.1 and Figures 2.10 and 2.12 where it can be seen that the estimated Laguerre model gives a very accurate representation of the true process.

	Noise-Free Case	Noise Case
p	0.13	0.13
Coefficients	$\hat{c}_1 = +3.1127 \times 10^{-3}$ $\hat{c}_2 = -2.3629 \times 10^{-2}$ $\hat{c}_3 = +7.0470 \times 10^{-2}$ $\hat{c}_4 = -9.9321 \times 10^{-2}$ $\hat{c}_5 = +5.7199 \times 10^{-2}$	$\hat{c}_1 = +2.6940 \times 10^{-3}$ $\hat{c}_2 = -2.4272 \times 10^{-2}$ $\hat{c}_3 = +6.9629 \times 10^{-2}$ $\hat{c}_4 = -1.0031 \times 10^{-1}$ $\hat{c}_5 = +5.6195 \times 10^{-2}$

Table 2.1: *Laguerre model coefficients for Example 2.5*

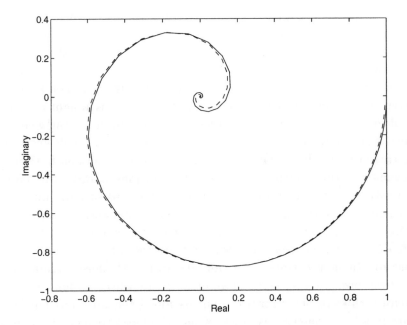

Figure 2.12: *Comparison of frequency responses for Example 2.5 (solid: true response; dashed: Laguerre model)*

2.6   A STRATEGY FOR IMPROVING THE LAGUERRE MODEL

The results of the previous section indicate that when the disturbance power spectrum is band-limited and does not significantly overlap the process frequency bandwidth, the estimated Laguerre model can be very accurate. However, the accuracy of the estimated Laguerre model will decrease if there is significant overlap. In practice, slow drifting disturbances are sometimes encountered during a step response test. The question is, can we do anything to improve the quality of the estimated Laguerre model in these situations? To answer this question, we examine the errors in the estimated Laguerre coefficients from a time domain perspective.

To look at the influence of disturbances on the estimated Laguerre coefficients in the time domain, we rewrite the estimate of the ith coefficient in Equation (2.72) as

$$\hat{c}_i = -\int_0^\infty l_i(t)g(t)dt - \int_0^\infty l_i(t)\xi(t)dt$$
$$= c_i - \Delta c_i \tag{2.101}$$

where c_i is the true Laguerre coefficient and Δc_i is the estimation error caused by the presence of the disturbance $\xi(t)$. Our objective is to reduce the value of Δc_i for a given disturbance in order to improve the accuracy of the estimates.

From Equation (2.101) we can see that Δc_i is an integration of the unknown disturbance $\xi(t)$ weighted by $l_i(t)$. How this weighting function behaves with respect to time determines the influence the disturbance has at different stages of the step response test on the estimated coefficient. Figures 2.13-2.15 illustrate the behaviour of the weighting functions $l_1(t)$ to $l_6(t)$, where the time scaling factor p has been chosen equal to unity. (When $p = 1$, the weighting functions can be regarded as functions of normalized time.) From these figures, it can be seen that the maximum absolute value of each weighting function occurs at $t = 0$, with the maximum value $|l_i(0)|$ equal to $p\sqrt{2p}(2(i - 1) + 1)$. The parameter p does not change the general shape of these weighting functions. However, the value of p does change the scale on the time domain axis and the amplitude of each $l_i(t)$. For instance, as p increases, the weighting functions shrink along the time axis but increase in amplitude. Conversely, as p decreases, the weighting functions expand along the time axis but their magnitudes decrease.

The weighting functions can be basically divided into three intervals with respect to the normalized time domain axis. The first time interval starts from $pt = 0$ and ends at the time that the weighting function makes

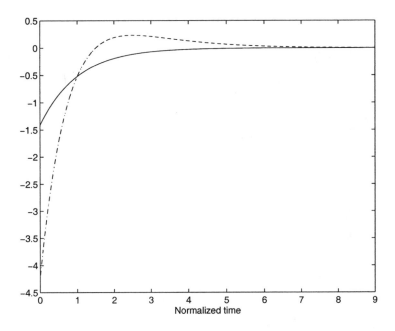

Figure 2.13: *Weighting functions in the time domain (solid: $\dot{l}_1(t)$; dash-dotted: $\dot{l}_2(t)$)*

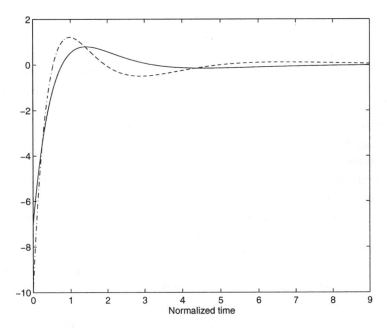

Figure 2.14: *Weighting functions in the time domain (solid: $\dot{l}_3(t)$; dash-dotted: $\dot{l}_4(t)$)*

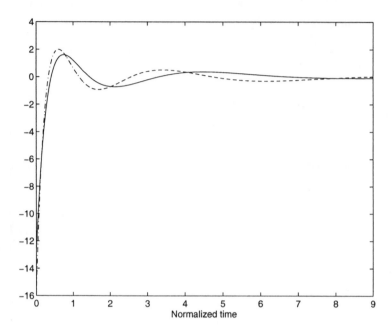

Figure 2.15: *Weighting functions in the time domain (solid: $\dot{l}_5(t)$; dash-dotted: $\dot{l}_6(t)$)*

its first zero crossing. The second time interval continues from there until the weighting function makes its final zero crossing. The final time interval extends from the last zero crossing to the time that the weighting function vanishes to zero.

During the first time interval, the weighting functions have relatively large magnitudes, and therefore disturbances at the beginning of the step response can be expected to lead to significant estimation errors. This is evident from Figures 2.16-2.18, where we show the products $\dot{l}_i(t) \times \xi(t)$, for $i = 1, 3, 6$, with $\xi(t)$ taken to be a normally distributed, white noise sequence with unit variance. As the index number i increases, these products become larger in magnitude, especially during the respective first time intervals.

During the second time interval, the weighting functions oscillate around zero but with smaller amplitudes as compared to their values in the first time interval. Therefore, the effect of disturbances during these time intervals can be expected to contribute less to the estimation errors. This can also be seen from Figures 2.16-2.18. In the final time interval, the weighting functions exponentially decay to zero and their amplitudes remain relatively small throughout the interval. Therefore, disturbances in the process step

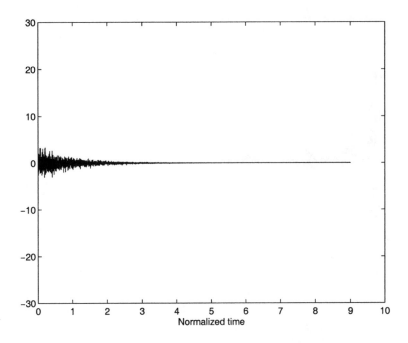

Figure 2.16: *Product of $\dot{l}_1(t)$ and $\xi(t)$*

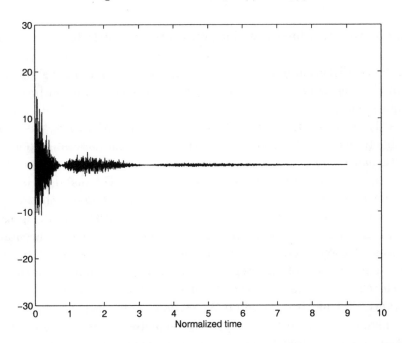

Figure 2.17: *Product of $\dot{l}_3(t)$ and $\xi(t)$*

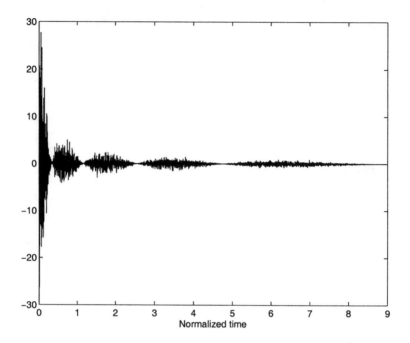

Figure 2.18: *Product of $\dot{l}_6(t)$ and $\xi(t)$*

response during this final time interval contribute little to the estimation errors.

Since the disturbances affecting the process output during the initial stage of the step response test make the greatest contribution to errors in the Laguerre coefficients, it would be most beneficial to eliminate the effect of these disturbances during this stage. However, in order to do this we need to be able to distinguish in some way between the true process response and the disturbance effect. Since many processes contain some amount of time delay, this separation can be performed to a degree. For instance, if there is some prior knowledge of at least a lower bound on the delay, the measured process output can be attributed completely to the disturbance up to this point in time. Therefore, a simple strategy from improving the accuracy of the estimated coefficients is to set the measured response identically equal to zero from the time of the step input until the lower bound on the delay has been reached. This will have the effect of eliminating the effect of the disturbances during this period on the estimation errors.

For high order processes, the process response is often close to zero for an initial period referred to as an apparent delay, and therefore the same procedure can be applied. For low order processes with no time delay, a

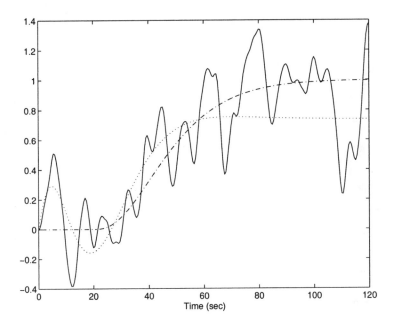

Figure 2.19: *Comparison of step responses for Example 2.6 (solid: measured response; dotted: Laguerre model; dash-dotted: true response)*

lower order Laguerre model is appropriate, in which case the disturbance effect on the estimates during the initial stage is less significant (see Figure 2.16). Regardless though for either case, eliminating the disturbance effect, even for just a short time period at the beginning of the step response, will improve the accuracy of the estimated Laguerre model.

Example 2.6. To illustrate the proposed data pretreatment strategy, we consider the step response of the process given in Equation (2.100) with the disturbance $\xi(t)$ generated by passing a white noise sequence of unit variance through a lowpass filter in the frequency range of $(0, 0.2\pi)$ radians/sec. The results of attempting to fit a $5th$ order Laguerre model using this data are shown in Figures 2.19 and 2.20, both in the time domain and the frequency domain. Without any data pretreatment, the algorithm obviously fails to produce a good model. However, if we have prior knowledge that the process has a time delay at least equal to or greater than 15 sec, and we set $\hat{g}(t) = 0$ for t in the interval $(0, 15)$, the Laguerre model fit is greatly improved as seen in Figures 2.21 and 2.22 and from the estimated coefficients in Table 2.2. It is important to note that this procedure does not require perfect knowledge of the delay, but only a lower bound on the delay.

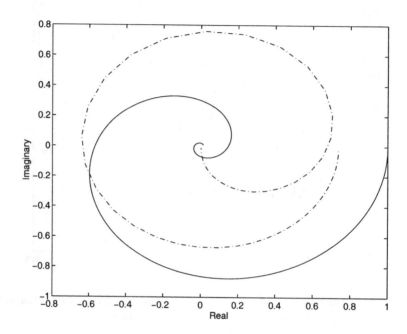

Figure 2.20: *Comparison of frequency responses for Example 2.6 (solid: true response; dash-dotted: Laguerre model)*

Noise-Free Case	No Pretreatment	With Pretreatment
$\hat{c}_1 = +3.1127 \times 10^{-3}$	$\hat{c}_1 = +7.1717 \times 10^{-2}$	$\hat{c}_1 = +6.1371 \times 10^{-3}$
$\hat{c}_2 = -2.3629 \times 10^{-2}$	$\hat{c}_2 = +1.1894 \times 10^{-1}$	$\hat{c}_2 = -2.9040 \times 10^{-2}$
$\hat{c}_3 = +7.0470 \times 10^{-2}$	$\hat{c}_3 = +1.8903 \times 10^{-1}$	$\hat{c}_3 = +7.0010 \times 10^{-2}$
$\hat{c}_4 = -9.9321 \times 10^{-2}$	$\hat{c}_4 = -9.8690 \times 10^{-2}$	$\hat{c}_4 = -1.0579 \times 10^{-1}$
$\hat{c}_5 = +5.7199 \times 10^{-2}$	$\hat{c}_5 = -5.2653 \times 10^{-2}$	$\hat{c}_5 = +5.5344 \times 10^{-2}$

Table 2.2: *Laguerre model coefficients for Example 2.6 (p = 0.13)*

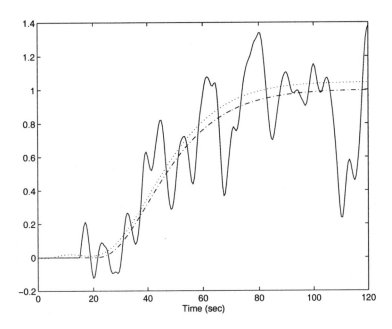

Figure 2.21: *Comparison of step responses using pretreated data for Example 2.6 (solid: pretreated data; dotted: Laguerre model; dash-dotted: true response)*

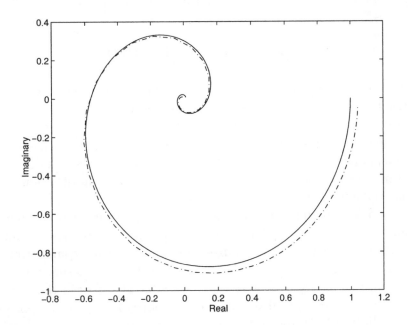

Figure 2.22: *Comparison of frequency responses using pretreated data for Example 2.6 (solid: true response; dash-dotted: Laguerre model)*

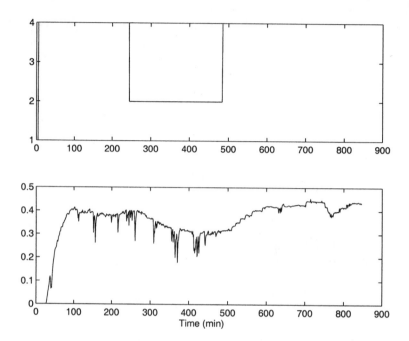

Figure 2.23: *Step response test data from polymer reactor. Upper diagram: initiator flow rate (ml/min); lower diagram: fraction monomer conversion*

2.7   MODELLING OF A POLYMER REACTOR

An automated pilot-scale 1-litre experimental polymer reactor system with facilities for on-line measurement of flow rate, temperature and density has been set up by Chien and Penlidis (1994*a*, *b*). These authors describe a set of open-loop process identification experiments and closed-loop control experiments performed on this system where monomer conversion is controlled in the presence of reactive impurities using the initiator flow rate as the manipulated variable.

Two open-loop step response tests were performed on this system by stepping the initiator pump feed flow rate down from 4 ml/min to 2 ml/min at $t = 243$ min (referred to as Step Response 1), and then stepping it up from 2 ml/min to 4 ml/min at $t = 484$ min (Step Response 2) as shown in Figure 2.23. From this figure, it is clear that several noise spikes in the measured monomer conversion appear during Step Response 1. The noise spikes that occurred during the initial part of the step response are handled using the proposed data pretreatment procedure. The spikes that occurred during the dynamic part of the monomer response are replaced using a simple linear interpretation. That is, suppose that two points along the monomer step response are deemed to be on either side of a noise spike denoted by $(t_1, \hat{g}(t_1))$ and $(t_2, \hat{g}(t_2))$, then the noise-free monomer response for $t_1 \leq t \leq t_2$ is given by

$$\hat{g}(t) = \frac{\hat{g}(t_1) - \hat{g}(t_2)}{t_1 - t_2} t - \frac{\hat{g}(t_1) - \hat{g}(t_2)}{t_1 - t_2} t_1 + \hat{g}(t_1) \qquad (2.102)$$

Since the polymer reactor is known to be nonlinear and there are noticeable differences between the two step responses, two separate Laguerre models are estimated from the data. The model order is chosen to be 3 for both cases.

Step Response 1

The settling time of the first step response is estimated to be approximately 200 sec and therefore the search interval for the optimal time scaling factor p is chosen to be $(\frac{4}{200}, \frac{20}{200})$. This interval is divided into 17 discrete values for computation of the Laguerre coefficients c_3 and c_4. The plot of $\hat{c}_3\hat{c}_4$ in Figure 2.24 shows that this product changes sign from positive to negative only once at $p = 0.043$. Using this p value, the resulting $3rd$ order Laguerre model is given by

$$G_1(s) = \frac{1.6943 \times 10^{-4} s^2 - 4.8114 \times 10^{-5} s + 3.3383 \times 10^{-6}}{(s + 0.043)^3} \qquad (2.103)$$

Figure 2.25 shows the comparison between the pretreated unit step response data and the step response generated from the estimated Laguerre model in Equation (2.103).

Step Response 2

Since both step responses have approximately the same settling time, the optimal pole location should lie in the same interval. For Step Response 2, the plot of $\hat{c}_3\hat{c}_4$ is shown in Figure 2.26 and indicates that the optimal time scaling factor is located at $p = 0.05$. The resulting $3rd$ order Laguerre model is given by

$$G_2(s) = \frac{-3.3388 \times 10^{-5} s^2 - 9.3278 \times 10^{-6} s + 6.7262 \times 10^{-6}}{(s + 0.05)^3} \qquad (2.104)$$

Figure 2.27 shows the comparison between the pretreated unit step response data and the step response generated from the estimated Laguerre model in Equation (2.104). The two unit step responses obtained from the Laguerre models in Equations (2.103) and (2.104) are compared in Figure 2.28, where it is clearly seen that there are some significant differences between the process dynamics that depend on the direction of the change in the initiator flow rate.

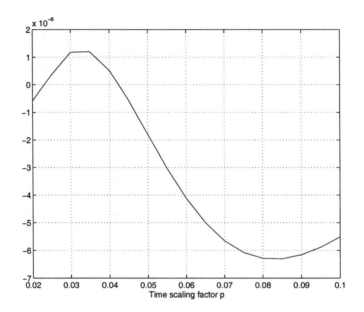

Figure 2.24: $\hat{c}_3\hat{c}_4$ *as a function of p for Step Response 1*

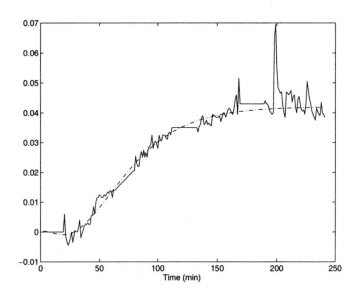

Figure 2.25: *Unit step response corresponding to Step Response 1 (solid: pretreated data; dash-dotted: Laguerre model)*

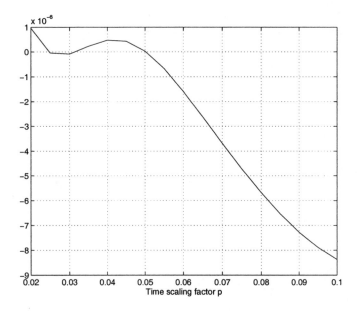

Figure 2.26: $\hat{c}_3\hat{c}_4$ *as a function of p for Step Response 2*

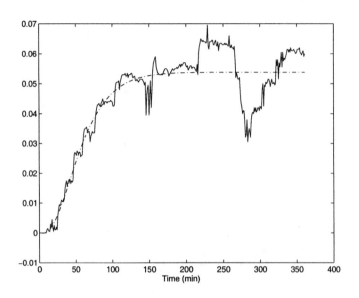

Figure 2.27: *Unit step response corresponding to Step Response 2 (solid: pretreated data; dash-dotted: Laguerre model)*

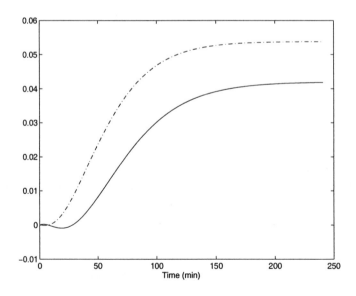

Figure 2.28: *Unit step response comparison (solid: G_1; dash-dotted: G_2)*

2.8   APPENDIX

Derivation of Covariance Matrix for Laguerre Model

Note that from Equations (2.72) and (2.75)

$$
\begin{bmatrix} \hat{c}_1 - c_1 \\ \hat{c}_2 - c_2 \\ \vdots \\ \hat{c}_N - c_N \end{bmatrix} = \begin{bmatrix} p & 0 & \cdots & 0 \\ 2p & p & \cdots & 0 \\ & \vdots & & \\ 2p & \cdots & 2p & p \end{bmatrix} \begin{bmatrix} \int_0^\infty \xi(t)l_1(t)dt \\ \int_0^\infty \xi(t)l_2(t)dt \\ \vdots \\ \int_0^\infty \xi(t)l_N(t)dt \end{bmatrix} \tag{2.105}
$$

Thus

$$
E[(\hat{C} - C)(\hat{C} - C)^T] = p^2 AQA^T \tag{2.106}
$$

where Q is an $N \times N$ symmetric matrix with its elements defined by

$$
Q_{ij} = E\left[\int_0^\infty \xi(t)l_i(t)dt \int_0^\infty \xi(t)l_j(t)dt\right] \tag{2.107}
$$

Since

$$
Q_{ij} = E\left[\int_0^\infty \int_0^\infty \xi(t)l_i(t)\xi(t')l_j(t')dtdt'\right] \tag{2.108}
$$

and letting $\tau = t - t'$, we obtain

$$
\begin{aligned}
Q_{ij} &= \int_0^\infty \int_{-t'}^{\infty-t'} l_i(t'+\tau)E[\xi(t'+\tau)\xi(t')]l_j(t')d\tau dt' & (2.109) \\
&= \int_0^\infty \int_{-t'}^{\infty-t'} l_i(t'+\tau)R_{\xi\xi}(\tau)l_j(t')d\tau dt' & (2.110) \\
&= \int_0^\infty \int_0^\infty l_i(s)R_{\xi\xi}(s-t')l_j(t')dsdt' \quad (s=t'+\tau) & (2.111) \\
&= \int_0^\infty l_j(t') \int_0^\infty l_i(s)R_{\xi\xi}(t'-s)dsdt' & (2.112)
\end{aligned}
$$

which gives the result in Equation (2.86). Let

$$
x(t') = \int_0^\infty R_{\xi\xi}(t'-s)l_i(s)ds \tag{2.113}
$$

Then, Equation (2.112) can be written as

$$
Q_{ij} = \int_0^\infty l_j(t')x(t')dt' \tag{2.114}
$$

and using Parseval's theorem

$$
Q_{ij} = \frac{1}{2\pi} \int_{-\infty}^\infty L_j^*(jw)X(jw)dw \tag{2.115}
$$

By substituting Equation (2.113) into the definition of the Fourier transform
we have

$$X(jw) = \int_{-\infty}^{\infty} x(t')e^{-jwt'}dt' \tag{2.116}$$

$$= \int_{-\infty}^{\infty}\int_{0}^{\infty} R_{\xi\xi}(t'-s)l_i(s)ds e^{-jwt'}dt' \tag{2.117}$$

$$= \int_{-\infty}^{\infty} R_{\xi\xi}(\tau)e^{-jw\tau}d\tau \int_{0}^{\infty} l_i(s)e^{-jws}ds \quad (\tau = t'-s)\tag{2.118}$$

$$= S_{\xi\xi}(w)L_i(jw) \tag{2.119}$$

Combining Equations (2.115) and (2.119) gives the result in Equation (2.87).

An equivalent result to Equation (2.85) can be obtained by taking the
Laplace transform of the derivatives of the Laguerre functions. It can be
readily shown that, from the state space representation of the Laguerre
network in Equation (2.19)

$$\begin{bmatrix} p & 0 & \cdots & 0 \\ 2p & p & \cdots & 0 \\ \vdots & & & \\ 2p & \cdots & 2p & p \end{bmatrix} \begin{bmatrix} L_1(jw) \\ L_2(jw) \\ \vdots \\ L_N(jw) \end{bmatrix} = \begin{bmatrix} \sqrt{2p} - jw \times L_1(jw) \\ \sqrt{2p} - jw \times L_2(jw) \\ \vdots \\ \sqrt{2p} - jw \times L_N(jw) \end{bmatrix} \tag{2.120}$$

By substituting Equation (2.120) into Equation (2.85), we obtain another
form of the covariance matrix for the coefficients as

$$E[(\hat{C}-C)(\hat{C}-C)^T] = \frac{1}{2\pi}\int_{-\infty}^{\infty} U(jw)S_{\xi\xi}(w)U^*(jw)dw \tag{2.121}$$

where

$$U(jw) = \begin{bmatrix} \sqrt{2p} - jw \times L_1(jw) \\ \sqrt{2p} - jw \times L_2(jw) \\ \vdots \\ \sqrt{2p} - jw \times L_N(jw) \end{bmatrix} \tag{2.122}$$

From Equation (2.121), the variance of the *ith* estimated coefficient is obtained as

$$Var[\hat{c}_i] = \frac{1}{2\pi}\int_{-\infty}^{\infty} |\sqrt{2p} - jw \times L_i(jw)|^2 S_{\xi\xi}(w)dw \tag{2.123}$$

Let $W_i(w) = 2p\left|1 - jw\frac{(jw-p)^{i-1}}{(jw+p)^i}\right|^2$. In order to derive Equation (2.89), we
will define $\theta = tan^{-1}\left(\frac{w}{p}\right)$, which leads to

$$
W_i(w) = 2p \left| 1 - (-1)^{i-1} \frac{w}{\sqrt{w^2 + p^2}} e^{j(\frac{\pi}{2} - (2(i-1)+1)\theta)} \right|^2 \tag{2.124}
$$

$$
= 2p \left[1 - (-1)^{i-1} \frac{2w}{\sqrt{w^2 + p^2}} sin[(2(i-1)+1)\theta] + \frac{w^2}{w^2 + p^2} \right]
$$

Equation (2.88) follows by noting that $sin\theta = \frac{w}{\sqrt{w^2 + p^2}}$ and that $W_i(w)$ is an even function.

Chapter 3

Least Squares and the PRESS Statistic using Orthogonal Decomposition

3.1 INTRODUCTION

The previous chapter discussed the problem of estimating a transfer function model from noisy step response data. Use of individual step tests to identify models for multivariable systems and systems with nonstationary disturbances is not always practical and may not lead to meaningful results. In these situations, input signals with multiple and perhaps random moves are required. In the next three chapters, new system identification tools are developed and used to build models for dynamic processes. This chapter discusses least squares parameter estimation using the orthogonal decomposition algorithm and proposes a simplified computational procedure for calculating the PRESS statistic using this algorithm. The PRESS is used extensively in this chapter and later chapters for structure selection of multivariable process models and disturbance models.

This chapter consists of six sections. Section 3.2 introduces the orthogonal decomposition algorithm proposed by Korenberg *et al.* (1988). Section 3.3 describes the concept of the *PRESS* statistic. Section 3.4 shows how to compute the *PRESS* using the orthogonal decomposition algorithm. Section 3.5 applies the *PRESS* statistic to the problem of model structure selection for dynamic systems. Section 3.6 shows how the *PRESS* statistic can be used for disturbance model structure selection.

Portions of this chapter have been reprinted from *Automatica* **32**, L. Wang and W.R. Cluett, "Use of PRESS residuals in dynamic system identification", pp. 781-784, 1996, with permission from Elsevier Science.

3.2   LEAST SQUARES AND ORTHOGONAL DECOMPOSITION

3.2.1   Least squares for dynamic models

The least squares method for parameter estimation is a central technique in the area of process identification. The method itself is particularly simple to apply if the selected model structure has the property of being linear-in-the-parameters. In this case, the least squares parameter estimates can be found analytically. For example, consider a model of the following form

$$y(k) = \phi_1(k)\theta_1 + \phi_2(k)\theta_2 + \cdots + \phi_n(k)\theta_n + \xi(k) \tag{3.1}$$

where $y(k)$ is the observed variable, $\theta_1, \theta_2, \ldots, \theta_n$ are the unknown parameters, $\phi_1(k), \phi_2(k), \ldots, \phi_n(k)$ are known functions and $\xi(k)$ is an error term. The variables $\phi_i(k)$ are called the regression variables or the regressors and the model in Equation (3.1) is called a linear regression model because it is linear with respect to its parameters.

For a dynamic system, we assume that the discrete-time process input sequence $\{u(k)\}$ and the discrete-time measured process output sequence $\{y(k)\}$, where $k = 1, 2, \ldots, M$ enumerates the sampling intervals, are related by the linear regression model given by

$$y(k) = \phi(k)^T \theta + \xi(k) \tag{3.2}$$

where $\theta^T = [\theta_1 \quad \theta_2 \quad \cdots \quad \theta_n]$ is the vector of process parameters, $\phi(k)^T = [\phi_1(k) \quad \phi_2(k) \quad \cdots \quad \phi_n(k)]$ is the regressor vector with each element $\phi_i(k)$ representing a linear or nonlinear function of present and past values of the process input and/or output, and $\xi(k)$ is the disturbance term. If the regressors are linear functions of the input and output, then we are assuming a linear model for the process. If the regressors contain nonlinear functions of the input and output, we are using a nonlinear model to represent the process. The problem is to determine the correct model structure and estimates of the model parameters such that the output predicted by the model

$$\hat{y}(k) = \phi(k)^T \hat{\theta} \tag{3.3}$$

is as close as possible to the measured output variable $y(k)$ in a least squares sense.

Equation (3.2) can also be written in a matrix form as

$$Y = \Phi\theta + \zeta \tag{3.4}$$

where $Y^T = [y(1) \ y(2) \ \cdots \ y(M)]$ contains the measured output values,

$$\Phi = \begin{bmatrix} \phi_1(1) & \phi_2(1) & \cdots & \phi_n(1) \\ \phi_1(2) & \phi_2(2) & \cdots & \phi_n(2) \\ \vdots & \vdots & & \vdots \\ \phi_1(M) & \phi_2(M) & \cdots & \phi_n(M) \end{bmatrix}$$

is the data matrix, and $\zeta^T = [\xi(1) \ \xi(2) \ \cdots \ \xi(M)]$ contains the disturbance sequence. The sum of squared prediction errors can then be written as

$$V = (Y - \Phi\theta)^T(Y - \Phi\theta) \tag{3.5}$$

and is minimized by the parameter vector $\hat{\theta}$ satisfying

$$\Phi^T\Phi\hat{\theta} = \Phi^TY \tag{3.6}$$

If the matrix $\Phi^T\Phi$ is invertible, the minimum of the least squares error is unique, with the estimated parameters given by

$$\hat{\theta} = (\Phi^T\Phi)^{-1}\Phi^TY \tag{3.7}$$

The matrix $\Phi^T\Phi$ is called the correlation matrix and the invertibility condition on this matrix is sometimes called the sufficient excitation condition for parameter estimation.

3.2.2 Orthogonal decomposition algorithm

The well known matrix decomposition theorem states that a positive square matrix P can be decomposed as

$$P = LDU \tag{3.8}$$

where L and U are unit lower and upper triangular matrices, and D is a diagonal matrix with all positive elements. If P is symmetric, then it can be shown that

$$L = U^T \tag{3.9}$$

and hence

$$P = U^TDU \tag{3.10}$$

Since the correlation matrix $\Phi^T\Phi$ is symmetric and positive definite under the assumption that it is invertible, it can be expressed as

$$\Phi^T\Phi = T^T W_d T \tag{3.11}$$

where T is a unit upper triangular matrix and W_d is a diagonal matrix with all positive elements.

To derive the least squares solution using the orthogonal decomposition result, we rewrite Equation (3.4) by inserting $T^{-1}T$

$$Y = \Phi(T^{-1}T)\theta + \zeta \tag{3.12}$$

Then, letting $W = \Phi T^{-1}$ and $g = T\theta$, we have

$$Y = Wg + \zeta \tag{3.13}$$

where g is the auxiliary model parameter vector and W is the transformed data matrix. Now we can show the connection between the matrix W (dimension $M \times n$) and the diagonal matrix W_d. From the definition of W

$$W^T W = (\Phi T^{-1})^T \Phi T^{-1} = (T^{-1})^T \Phi^T \Phi T^{-1} \tag{3.14}$$

Noting that $\Phi^T\Phi$ can be decomposed into $T^T W_d T$, then

$$W^T W = W_d \tag{3.15}$$

Since W_d is a diagonal matrix, W is an orthogonal matrix.

Taking advantage of the orthogonality of the matrix W, the least squares problem can be solved in terms of the auxiliary parameter vector g and the transformed data matrix W. That is, based on Equation (3.13), the vector g can be estimated from the least squares solution as

$$\hat{g} = W_d^{-1} W^T Y \tag{3.16}$$

which minimizes the loss function

$$V = (Y - Wg)^T (Y - Wg) \tag{3.17}$$

Since W_d is a diagonal matrix, Equation (3.16) does not require matrix inversion and can be solved on an element-by-element basis. The least squares estimate of the original parameter vector $\hat{\theta}$ is then obtained using the relation

$$\hat{\theta} = T^{-1}\hat{g} \tag{3.18}$$

Because T is a unit upper triangular matrix, it is numerically well conditioned, and therefore its inversion is straightforward.

3.3 THE *PRESS* STATISTIC

The conventional residuals from the least squares estimator in Equation (3.7) are defined as

$$
\begin{aligned}
e(k) &= y(k) - \phi(k)^T \hat{\theta} \\
&= y(k) - \hat{y}(k)
\end{aligned}
\tag{3.19}
$$

These residuals are measures of the quality of the model fit for the given data set, but do not assess the predictive capability of the model. Note that because both the output $y(k)$ and the regressor $\phi(k)$ are used to estimate θ, $\hat{y}(k)$ is not independent of $y(k)$. In fact, the least squares procedure is designed to produce properties that will result in residuals $e(k)$ that are smaller than the true prediction errors (Myers, 1990).

In order to avoid the correlation that exists between the conventional residuals and process output data, it has been suggested in the dynamic system identification literature that the data be split into an estimation set, which is used to estimate the parameters, and a testing set, which is used to judge the predictive capability of the fitted model. The residuals associated with the testing set may be used for model structure determination and are referred to as the true prediction errors because, in this case, $y(k)$ and $\hat{y}(k)$ are independent. This approach is useful for revealing the structure of a dynamic system subject to disturbances where it is believed that the disturbance sequence will never be exactly duplicated from the estimation set to the testing set.

The use of a new data set or data splitting for the purpose of cross validation may not always be applicable or desirable. In addition, the results depend on the location of the split. An alternative is to define the prediction error as

$$
\begin{aligned}
e_{-k}(k) &= y(k) - \phi(k)^T \hat{\theta}_{-k} \\
&= y(k) - \hat{y}_{-k}(k)
\end{aligned}
\tag{3.20}
$$

where $e_{-k}(k)$, $k = 1, 2, \ldots, M$, are called the *PRESS* residuals and $\hat{\theta}_{-k}$ has been estimated using the least squares algorithm without including $\phi(k)$ and $y(k)$. The *PRESS* residuals $e_{-k}(k)$ represent the true prediction errors as $y(k)$ and $\hat{y}_{-k}(k)$ are independent. In the same spirit as data splitting, the *PRESS* residuals give us information in the form of M cross validations in which the fitting sample for each cross validation is of size $M - 1$.

It has been shown in the literature (see e.g. Myers, 1990) that through use of the Sherman-Morrison-Woodbury theorem, the *PRESS* residuals

$e_{-k}(k)$ can be calculated according to the following equation

$$e_{-k}(k) = \frac{e(k)}{1 - \phi(k)^T (\Phi^T \Phi)^{-1} \phi(k)} \qquad (3.21)$$

Then the P̲R̲ediction E̲rror S̲um of S̲quares ($PRESS$) statistic is defined as

$$PRESS = \sum_{k=1}^{M} e_{-k}(k)^2 \qquad (3.22)$$

and the average $PRESS$ as

$$PRESS_{ave} = \sqrt{\frac{\sum_{k=1}^{M} e_{-k}(k)^2}{M-1}} \qquad (3.23)$$

Both the $PRESS$ and $PRESS_{ave}$ provide measures of the predictive capability of the estimated model.

3.4   COMPUTATION OF THE *PRESS* STATISTIC

Computation of the true prediction errors $e_{-k}(k)$, $k = 1, 2, \ldots, M$, is a tremendous task in dynamic system identification where we typically face a large amount of data (M) and possibly high dimensionality (n) of the parameter vector θ. It will be shown here that by using the orthogonal decomposition algorithm, the computation of the $PRESS$ residuals is simplified to an extent that its calculation can be viewed as a byproduct of the algorithm. The following theorem presents the cornerstone for computation of the $PRESS$ statistic.

Theorem 3.1: Let $w_i(.)$ denote the *ith* column of W and $\hat{g}_i$ represent the *ith* estimated auxiliary parameter. Then the $PRESS$ residuals $e_{-k}(k)$, $k = 1, 2, \ldots, M$, for the original model with n parameters are given by

$$e_{-k}(k) = \frac{y(k) - \sum_{i=1}^{n} w_i(k)\hat{g}_i}{1 - \sum_{i=1}^{n} \frac{w_i(k)^2}{\|w_i\|^2}} \qquad (3.24)$$

where $\|w_i\| = \sqrt{\sum_{k=1}^{M} w_i(k)^2}$ is the norm of w_i.

Proof: From Equation (3.13), we can write the conventional residuals in terms of the orthogonalized data matrix and the auxiliary parameter estimates

$$e(k) = y(k) - \sum_{i=1}^{n} w_i(k)\hat{g}_i \qquad (3.25)$$

We also note that from the definitions of $\phi(k)$ and Φ in Equations (3.2) and (3.4) we have

$$\phi(k)^T(\Phi^T\Phi)^{-1}\phi(k) = diag_k(\Phi(\Phi^T\Phi)^{-1}\Phi^T) \tag{3.26}$$

From the definition of W, we obtain $\Phi = WT$ which gives

$$\begin{aligned}\Phi(\Phi^T\Phi)^{-1}\Phi^T &= WT(T^TW^TWT)^{-1}T^TW^T\\ &= W(W^TW)^{-1}W^T\end{aligned} \tag{3.27}$$

Hence

$$\begin{aligned}\phi(k)^T(\Phi^T\Phi)^{-1}\phi(k) &= diag_k(W(W^TW)^{-1}W^T)\\ &= \sum_{i=1}^{n}\frac{w_i(k)^2}{||w_i||^2}\end{aligned} \tag{3.28}$$

From the expression for the *PRESS* residuals $e_{-k}(k)$ in Equation (3.21), the result in Equation (3.24) follows.

Remarks:

- It is seen from Equations (3.15), (3.16) and (3.17) that the sum of squares of the conventional residuals for a model with n parameters is given by

$$\begin{aligned}V_n &= (Y - W\hat{g})^T(Y - W\hat{g})\\ &= Y^TY - \hat{g}^TW^TY - Y^TW\hat{g} + \hat{g}^TW^TW\hat{g}\\ &= Y^TY - \hat{g}^TW^TY - Y^TW\hat{g} + \hat{g}^TW_dW_d^{-1}W^TY\\ &= \sum_{k=1}^{M}y(k)^2 - \sum_{i=1}^{n}\hat{g}_i^2||w_i||^2\end{aligned} \tag{3.29}$$

Therefore, for a model of order $n + 1$

$$V_n - V_{n+1} = \hat{g}_{n+1}^2||w_{n+1}||^2 \tag{3.30}$$

which shows that the sum of squares of the conventional residuals is nonincreasing with respect to model order. However, the *PRESS* statistic defined in Equation (3.22) does not always decrease as more terms are added to the model. In fact, if a term is added to the model and the *PRESS* increases, this indicates that the predictive capability of the model is better without that term.

- By examining Equations (3.24) and (3.25), it can be seen that the true prediction errors $e_{-k}(k)$ are in fact a weighted version of the conventional residuals $e(k)$. The weighting factor $(1 - \sum_{i=1}^{n} \frac{w_i(k)^2}{||w_i||^2})^{-1}$ gives large weights to conventional residuals associated with data points where prediction is poor.

- The computation of the *PRESS* residuals $e_{-k}(k)$ using Equation (3.24) only requires the orthogonal matrix W and the auxiliary parameter vector $\hat{g}$. Hence, the value of the *PRESS* can be used to detect the significance of each additional term in the original model without actually having to compute $\hat{\theta}$. This is a result of the structure of the T matrix and the relationship between $\hat{g}$ and $\hat{\theta}$.

3.5   USE OF *PRESS* FOR PROCESS MODEL SELECTION

This section illustrates application of the *PRESS* statistic for process model structure selection. Ljung (1987) used data collected from a laboratory-scale Process Trainer to illustrate various identification techniques and examined the sum of squared conventional residuals and Akaike's information theoretic criterion (AIC) for model structure selection. Two sets of input-output data collected from this process are available within MATLAB. We use the entire first set of data ($M = 1000$), called DRYER2 in MATLAB, for this study. Two different model structures are examined here, namely the ARX and FIR model structures, with the objective to find the model within a particular structure that produces the smallest *PRESS*.

ARX Models

For a linear, time invariant system, the regressor associated with the ARX model is chosen to have the following form

$$\phi(k)^T = [-y(k-1) \quad u(k-d-1) \quad -y(k-2) \quad u(k-d-2) \quad \cdots \quad \cdots \quad] \quad (3.31)$$

where d is the time delay of the process expressed as an integer multiple of the sampling interval. This arrangement ensures that an increasing number of terms in the regressor corresponds to an increase in the model order. For completeness, the process output $y(k-i)$ and the process input $u(k-d-i)$ always appear in pairs. The process output $y(k)$ is assumed to have the following form

$$y(k) = \phi(k)^T \theta + \xi(k) \quad (3.32)$$

model order n_l	time delay d	$PRESS$
1	1	29.999
2	1	2.7895
3	1	1.5734
14 (best)	**1**	**1.3956**
1	2	13.947
2	2	1.6609
3	2	1.4895
13 (best)	**2**	**1.4318**
1	3	10.598
2 (best)	**3**	**7.6355**
3	3	7.6639

Table 3.1: *PRESS values for ARX models using DRYER2 data set*

where $\phi(k)^T = [-y(k-1) \ u(k-d-1) \ \ldots \ \ldots \ -y(k-n_l) \ u(k-d-n_l)]$.
For this model, the time delay d and model order n_l must be determined in
addition to estimation of the parameter vector θ.

In order to determine the time delay d, we have calculated the $PRESS$
statistic using Equation (3.22) for $d = 1, 2$ and 3 over a range of model
orders, as shown in Table 3.1. It can be seen that for $d = 1$ and 2, the
$PRESS$ values are similar for $n_l \geq 3$. However, for $d = 3$, the minimum
$PRESS$ value corresponding to the best model order is significantly larger
than the minimum $PRESS$ for $d = 1$ and 2. Therefore, we can conclude
that the time delay d for this system is either one or two sampling intervals
which agrees with the results presented in Ljung (1987).

From the results presented in Table 3.1, our conclusion is that the best
ARX model, in terms of predictive capability, for the Process Trainer is
either a 14*th* order model (28 model terms) with $d = 1$ or a 13*th* order
model (26 model terms) with $d = 2$. The reason such high order models
were selected is that, with low order ARX models, there must exist a mis-
match between the assumed and actual noise structures. Evidence for this
statement can be found in Ljung (1987) where the addition of a noise model
was found to give improvement in terms of the AIC. In a similar situation,
Kosut and Anderson (1994) have fit least squares ARX models using cross
validation for model order selection and have found that high order ARX
models are often necessary.

If the model structure must be restricted, the above results also indicate

M	best n_l ($PRESS$)	$PRESS_{ave}$	best n_l (FPE)	FPE
200	3	4.18×10^{-2}	7	1.57×10^{-3}
400	14	4.10×10^{-2}	17	1.59×10^{-3}
600	14	3.84×10^{-2}	7	1.42×10^{-3}
800	11	3.81×10^{-2}	13	1.41×10^{-3}
1000	13	3.79×10^{-2}	13	1.40×10^{-3}

Table 3.2: *Comparison between PRESS and FPE using different data lengths from DRYER2 data set*

that a much lower order ARX model could be used. Note that for $d = 2$, the change in the $PRESS$ value going from $n_l = 3$ (1.4895) to $n_l = 13$ (1.4318) is small and therefore we could have selected $n_l = 3$ without a significant decrease in the predictive capability of the model.

In order to examine the consistency and robustness of the $PRESS$ statistic as a method for model structure selection and to compare it with a standard selection criterion, the best ARX model order for the Process Trainer has been determined using different amounts of data ($M = 200$, 400, 600, 800, 1000) from the original data set (DRYER2). The best model order using the $PRESS$ statistic was chosen based on the minimum value of the average prediction error $PRESS_{ave}$, which incorporates the data length effect. For comparison, the best model order has also been determined using the minimum value of Akaike's Final Prediction Error (FPE) criterion (Ljung, 1987) as calculated by MATLAB. A value of $d = 2$ was used throughout this comparison.

Table 3.2 presents a summary of our findings. From this table, it can be seen that the $PRESS$ statistic selects a model order which is close to the best order obtained using the complete data set ($M = 1000$) starting with $M = 400$. However, the FPE criterion does not select a model order close to the best order until $M = 800$. In addition, it is interesting to note that with the smallest data set examined ($M = 200$), the $PRESS$ statistic also manages to select the best low order model ($n_l = 3$), according to our earlier analysis, but this was not the case for the FPE criterion. These results show that, for this example, the $PRESS$ statistic provides a consistent order estimate and is more robust than the FPE criterion in terms of sensitivity to data length effects.

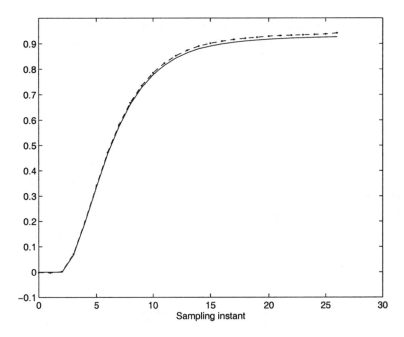

Figure 3.1: *Estimated step responses for Process Trainer (solid: ARX model; dashed: FIR model)*

FIR Models

The regressor for the FIR model structure is of the form

$$\phi(k)^T = [u(k-1) \;\; u(k-2) \;\; u(k-3) \;\; \dots \;\; u(k-N)]$$

and the model parameters correspond to the discrete-time unit impulse response coefficients of the process. Note that estimation of the process time delay is not required in this case.

The best FIR model, in terms of predictive capability, for the Process Trainer is a $27th$ order model ($N = 27$) corresponding to a *PRESS* of 9.13. Figure 3.1 shows the step response of the best FIR model and the step response of the best ARX model with $d = 2$ where it can be seen that the estimated step responses from the two models are almost identical.

3.6   USE OF *PRESS* FOR DISTURBANCE MODEL SELECTION

In the areas of process identification (Ljung, 1987), optimal stochastic controller design (Box and Jenkins, 1976) and controller performance assessment (Harris, 1989), it is often desirable or necessary to determine a model

from a single time series. The idea is to represent the time series, $\xi(k)$, by a
zero mean, white noise sequence, $\epsilon(k)$, filtered through a transfer function

$$\xi(k) = \frac{E(z)}{D(z)}\epsilon(k) \qquad (3.33)$$

where $E(z)$ and $D(z)$ are polynomials in the backward shift operator z^{-1}.
$E(z)$ is assumed to have its roots with respect to z strictly inside the unit
circle. If $\xi(k)$ represents a disturbance term, we refer to Equation (3.33) as
a disturbance model. Since $\epsilon(k)$ is not measurable, the direct estimation of
$E(z)$ and the denominator $D(z)$ based only on $\xi(k)$ is, in general, a nonlinear
parameter estimation problem (Box and Jenkins, 1976). Here, we prefer to
formulate this as a linear parameter estimation problem.

We begin by replacing $\frac{E(z)}{D(z)}$ by another transfer function of the form $\frac{1}{F(z)}$.
With the assumption that $E(z)$ is stable, we can write

$$\frac{E(z)}{D(z)} = \frac{1}{\frac{D(z)}{E(z)}} \approx \frac{1}{F(z)} \qquad (3.34)$$

where $F(z) = 1 + f_1 z^{-1} + f_2 z^{-2} + \cdots + f_m z^{-m}$, for some finite value of m.
Then, Equation (3.33) can be written in the following linear regression form

$$\begin{aligned}
\xi(k) &= -f_1\xi(k-1) - f_2\xi(k-2) - \cdots - f_m\xi(k-m) + \epsilon(k) \\
\xi(k) &= \phi_f(k)^T\theta_f + \epsilon(k)
\end{aligned} \qquad (3.35)$$

where

$$\phi_f(k)^T = [-\xi(k-1) \quad -\xi(k-2) \quad \ldots \quad -\xi(k-m)]$$

and θ_f is the corresponding parameter vector containing $f_1, f_2, \ldots, f_m$. The
parameters of the model in Equation (3.35) can now be estimated analytically using a least squares estimator based on the sequence $\xi(k)$. However, it
is important to choose the correct model order m to ensure that an adequate
approximation in Equation (3.34) has been achieved. This can be done using a variety of measures such as the AIC, or by performing a whiteness test
on the model residuals.

The least squares estimator and the *PRESS* statistic are used here
as a new way to estimate the best disturbance model in Equation (3.35).
This is an ideal application because the objective is to choose the most
parsimonious model (smallest m) while, at the same time, achieving a good
approximation in Equation (3.34). An indication that a sufficient model
order has been chosen is whether the residuals associated with the model

behave like white noise. Recall that the *PRESS* residuals give information in the form of M cross validations in which the fitting sample for each cross validation is of size $M - 1$. When the residuals become white for some model order m, there is no correlation structure remaining in the residuals. The *PRESS* statistic will indicate when this has been achieved by either resulting in increasing *PRESS* values or *PRESS* values that fluctuate around a constant value, for model orders greater than m. Our own rules for determining the most appropriate disturbance model order m are based on whichever of the following conditions is satisfied first:

(a) select the model order to be m if the *PRESS* value associated with a model of order $m + 1$ $(PRESS(m + 1))$ is greater than the *PRESS* associated with a model of order m $(PRESS(m))$;

(b) select the model order to be m if $\frac{PRESS(m) - PRESS(m+1)}{PRESS(m)} < 0.002$.

The following simulation examples are used to illustrate the application of this approach to disturbance modelling.

Example 3.1. Consider the disturbance model

$$\frac{E(z)}{D(z)} = \frac{0.3}{(1 - 0.7z^{-1})(1 - 0.8z^{-1})(1 - 0.9z^{-1})} \tag{3.36}$$

$$= \frac{0.3}{1 - 2.4z^{-1} + 1.91z^{-2} - 0.504z^{-3}} \tag{3.37}$$

which has the exact form of $\frac{1}{F(z)}$ with $m = 3$. For this simulation, the source of the disturbance $\epsilon(k)$ is chosen to be a zero mean, normally distributed, white noise sequence with unit variance. The disturbance $\xi(k)$ is generated by filtering $\epsilon(k)$ through the disturbance model. To simulate the stochastic nature of the problem, $\epsilon(k)$ is realized in MATLAB using seeds 1 through 1000 in order to generate 1000 different realizations of the disturbance sequence. Out of the 1000 simulation experiments, when the model order was chosen based on condition (a), m was selected to be 3 in 640 cases and 4 in 229 cases. When the model order was chosen based on condition (b), the model order was selected to be 3 in 923 cases and 4 in 73 cases.

For illustrative purposes, we examine a single experiment using seed number 888. Figure 3.2 shows the behaviour of the *PRESS* with respect to the disturbance model order m. This figure shows that the *PRESS* fluctuates somewhat randomly after the model order has reached 3, and eventually increases at higher model orders. The estimated disturbance model in this

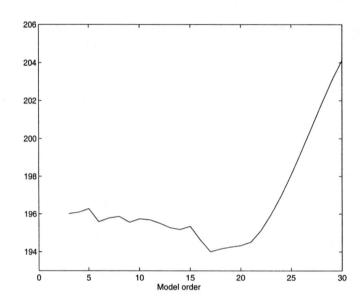

Figure 3.2: *Behaviour of the PRESS for Example 3.1 (seed 888)*

case is $\hat{F}(z) = 1 - 2.4537z^{-1} + 2.0074z^{-2} - 0.5535z^{-3}$. The autocorrelation function of the conventional residuals associated with this model is shown in Figure 3.3 with 95% confidence intervals. This figure illustrates an autocorrelation function that is well inside the confidence intervals after the zero*th* lag, indicating that the residuals associated with this model are behaving like white noise.

Example 3.2. Consider the disturbance model

$$\frac{E(z)}{D(z)} = \frac{0.3(1 + 0.6z^{-1})(1 + 0.3z^{-1})}{(1 - 0.7z^{-1})(1 - 0.8z^{-1})(1 - 0.9z^{-1})} \qquad (3.38)$$

Again, 1000 realizations of the disturbance sequence $\xi(k)$ have been generated by filtering a zero mean, normally distributed, white noise sequence $\epsilon(k)$ with unit variance through the disturbance model. Here the assumed model structure $\frac{1}{F(z)}$ is only an approximation the true disturbance model. For this example, we applied our rule for determining the disturbance model order, i.e. choose m based on whichever condition (a) or (b) is satisfied first. Out of the 1000 experiments, a model order of 7 was chosen in 635 cases, a model order of 8 in 135 cases, and a model order of 9 in 229 cases.

For illustrative purposes, we examine a single experiment using seed number 369. Figure 3.4 shows the behaviour of the *PRESS*. When the

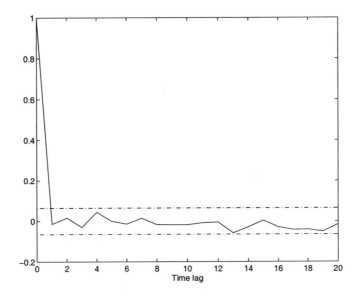

Figure 3.3: *Autocorrelation function of conventional residuals for Example 3.1 (seed 888)*

model order increases from 7 to 8, condition (b) determines the model order of $m = 7$. The estimated disturbance model is

$$
\begin{aligned}
\hat{F}(z) &= 1 - 3.3252z^{-1} + 4.7244z^{-2} - 4.0529z^{-3} \\
&+ 2.6555z^{-4} - 1.5406z^{-5} + 0.7164z^{-6} - 0.1775z^{-7} \quad (3.39)
\end{aligned}
$$

The autocorrelation function of the conventional residuals associated with this model is shown in Figure 3.5 where it can be seen that the residuals are behaving like white noise.

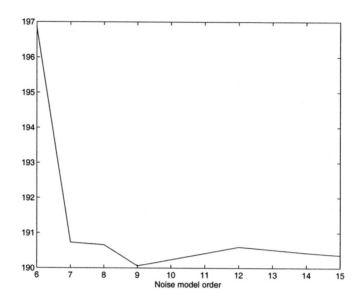

Figure 3.4: *Behaviour of PRESS for Example 3.2 (seed 369)*

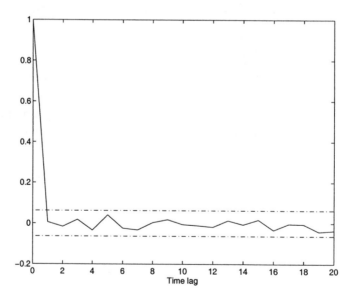

Figure 3.5: *Autocorrelation function of conventional residuals for Example 3.2 (seed 369)*

Chapter 4

Frequency Sampling Filters in Process Identification

4.1 INTRODUCTION

The next two chapters focus on process identification using the frequency sampling filter (FSF) model structure. The FSF model is obtained from a linear transformation of the FIR model and, as a result, inherits the main advantage of the FIR model, namely that the only prior information required from the user is an estimate of the process settling time. It is shown in this chapter that the transformation from the time domain FIR model to the frequency domain FSF model permits a reduction in the number of significant model parameters. It is also shown that when using a least squares algorithm to estimate the parameters of the FSF model, the associated correlation matrix has elements that are directly related to the energy content of the input signal used in the identification experiment.

This chapter consists of six sections. Section 4.2 introduces the FSF model structure. Section 4.3 examines the properties of the FSF model with a fast data sampling rate. Section 4.4 introduces the concept of a reduced order FSF model. Section 4.5 discusses the use of least squares for estimating the FSF model parameters from input-output data. Section 4.6 examines the nature of the correlation matrix that arises when using a least squares estimator with an FSF model and the relationship between the elements of this matrix and the energy content of the input signal.

Portions of this chapter have been reprinted from *Automatica* **33**, L. Wang and W.R. Cluett, "Frequency-sampling filters: an improved model structure for step-response identification", pp. 939-944, 1997, with permission from Elsevier Science.

4.2 THE FREQUENCY SAMPLING FILTER MODEL

We begin with the single input, single output (SISO) case and assume that the process to be identified is stable, linear, time invariant and can be represented by the following discrete-time, finite impulse response (FIR) transfer function model

$$G(z) = \sum_{i=0}^{N-1} h_i z^{-i} \tag{4.1}$$

where N is the model order chosen such that the FIR model coefficients $h_i \approx 0$ for all $i \geq N$, and z^{-1} is the backward shift operator. The model order N can be determined from an estimate of the process settling time T_s, where $N = \frac{T_s}{\Delta t}$ and Δt is the sampling interval. The FIR model is widely used in the field of process identification because it requires no prior knowledge about the process other than its settling time T_s.

To derive the frequency sampling filter (FSF) model, we first make use of the inverse Discrete Fourier Transform (DFT) relationship between the process frequency response and its impulse response, under the assumption that N is an odd number

$$h_i = \frac{1}{N} \sum_{l=-\frac{N-1}{2}}^{\frac{N-1}{2}} G(e^{j\frac{2\pi l}{N}}) e^{j\frac{2\pi l i}{N}} \tag{4.2}$$

This relationship maps a set of discrete-time frequency response coefficients, $G(e^{j\frac{2\pi l}{N}})$, $l = 0, \pm 1, \pm 2, \ldots, \pm \frac{N-1}{2}$ into the set of discrete-time unit impulse response coefficients, h_i, $i = 0, \ldots, N-1$. Substituting Equation (4.2) into Equation (4.1) gives

$$G(z) = \sum_{i=0}^{N-1} \frac{1}{N} \sum_{l=-\frac{N-1}{2}}^{\frac{N-1}{2}} G(e^{j\frac{2\pi l}{N}}) e^{j\frac{2\pi l i}{N}} z^{-i} \tag{4.3}$$

Interchanging the summations in Equation (4.3) gives the transfer function in its frequency sampling filter model form

$$G(z) = \sum_{l=-\frac{N-1}{2}}^{\frac{N-1}{2}} G(e^{j\frac{2\pi l}{N}}) \frac{1}{N} \frac{1 - z^{-N}}{1 - e^{j\frac{2\pi l}{N}} z^{-1}} \tag{4.4}$$

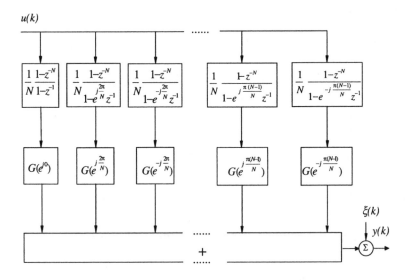

Figure 4.1: *Block diagram of frequency sampling filter model structure*

where we have used the result

$$\sum_{i=0}^{N-1} e^{j\frac{2\pi l i}{N}} z^{-i} = \frac{1 - z^{-N}}{1 - e^{j\frac{2\pi l}{N}} z^{-1}} \qquad (4.5)$$

We define the set of transfer functions found in Equation (4.4)

$$H^l(z) = \frac{1}{N} \frac{1 - z^{-N}}{1 - e^{j\frac{2\pi l}{N}} z^{-1}} \qquad (4.6)$$

for $l = 0, \pm 1, \pm 2, \ldots, \pm \frac{N-1}{2}$, as the frequency sampling filters and we will refer to $\frac{2\pi l}{N}$ radians as the centre frequency of the lth filter, $H^l(z)$.

Figure 4.1 shows a block diagram of the frequency sampling filter model being used to represent the process, where $u(k)$ is the discrete-time process input, $y(k)$ is the discrete-time measured process output and $\xi(k)$ is the disturbance. In the figure, it is shown that the process input first passes through the set of frequency sampling filters arranged in parallel. Then, the output of each filter is weighted by the discrete-time process frequency response evaluated at the corresponding centre frequency. Finally, the weighted filter

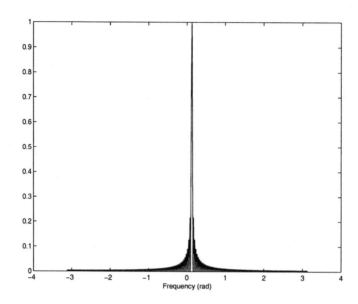

Figure 4.2: *Magnitude of the frequency response of the FSF filter with centre frequency at $\frac{2\pi l}{N}$ (l = 4 and N = 201)*

outputs are summed to form the noise-free process output. The FSF filters are narrow band-limited around their respective centre frequencies (see Figure 4.2). All the filters have identical frequency responses except for the location of their centre frequencies.

Since the FSF model is obtained from a linear transformation of the FIR model, as shown in Equations (4.1)-(4.4), it shares some features with the FIR model. For instance, the FSF model only requires prior information about the process settling time expressed in terms of N, and the number of unknown parameters in the FSF model is equal to the number of unknown parameters in the FIR model (N). However, there are two major differences between these models. First, the parameters of the FSF model correspond to the discrete-time frequency response coefficients, while the parameters of the FIR model correspond to the discrete-time unit impulse response coefficients. Second, with the FSF model, the elements of the regressor vector to be used later in this chapter for estimating the frequency response coefficients are formed by passing the process input through the set of narrow band-limited frequency sampling filters. On the other hand, with the FIR model, the regressor vector used for estimating the impulse response coefficients contains simple delayed values of the process input. These two major differences give the FSF model some key advantages over the FIR model from a process identification perspective.

One other point worth making is the relationship between the FSF model parameters and the entire process frequency response. As stated earlier, the parameters of the FSF model are the values of the discrete-time process frequency response at $w = w_l = \frac{2\pi l}{N}$ radians, where $l = 0, \pm 1, \pm 2, \ldots, \pm \frac{N-1}{2}$. Other process frequency response information can be readily obtained using the FSF model by letting $z = e^{jw}$ and evaluating

$$G(e^{jw}) = \sum_{l=-\frac{N-1}{2}}^{\frac{N-1}{2}} G(e^{j\frac{2\pi l}{N}}) \frac{1}{N} \frac{1 - e^{-jwN}}{1 - e^{j\frac{2\pi l}{N}} e^{-jw}} \qquad (4.7)$$

When $w = w_l$, it can be shown that $H^i(e^{jw}) = 0$ for $i \neq l$ and $H^i(e^{jw}) = 1$ for $i = l$, where i is an integer like l in the range $[-\frac{N-1}{2}, \frac{N-1}{2}]$. In this case, the value of the process frequency response in Equation (4.7) reduces to the value of the process frequency response coefficient $G(e^{j\frac{2\pi l}{N}})$.

4.3   PROPERTIES OF THE FSF MODEL WITH FAST SAMPLING

It is well known that many discrete-time models obtained with a fast sampling rate have some undesirable properties. For example, the order of the FIR model in Equation (4.1) is inversely proportional to the sampling interval, i.e. $N \to \infty$ as the sampling interval $\Delta t \to 0$. With rational transfer function models, it was shown by Åström *et al.* (1984) that a continuous-time process with no zeros or with zeros in the left half plane will often give rise to a discrete-time model having zeros outside the unit circle as the sampling period tends to zero. In addition, the poles of the discrete-time model will tend to the unit circle. By comparison, we will show here that the FSF model shares one of the major advantages of delta operator models (Middleton and Goodwin, 1990) in that the FSF model parameters converge to their continuous-time counterparts as $\Delta t \to 0$.

The properties of the FSF model with respect to choice of the sampling interval are summarized in the theorem below.

Theorem 4.1. We assume that:

- the underlying continuous-time process is stable, linear and time invariant with Laplace transfer function $G(s)$;

- the underlying continuous-time process has a unit impulse response $h(t)$ with finite settling time T_s such that for $t \geq T_s$, $h(t) \approx 0$.

We set the parameter N in the FSF model given by Equation (4.4) as $N = \frac{T_s}{\Delta t}$ for a given sampling interval $\Delta t > 0$. Then, as $\Delta t \to 0$, the parameters of the FSF model, $G(e^{j\frac{2\pi l}{N}})$, converge to $G(jw_l)$ at $w_l = \frac{2\pi l}{T_s}$ radians/time.

Proof: Taking the Fourier transform of $h(t)$ gives the continuous-time frequency response of the process as

$$
\begin{aligned}
G(jw) &= \int_0^{\infty} h(t)e^{-jwt}dt \\
&\approx \int_0^{T_s} h(t)e^{-jwt}dt \\
&= \lim_{\Delta t \to 0} \sum_{i=0}^{N-1} h(i\Delta t)e^{-jwi\Delta t}\Delta t
\end{aligned}
\tag{4.8}
$$

If we choose to evaluate the frequency response up to the Nyquist frequency, $\frac{\pi}{\Delta t}$ radians/time, in increments of $\frac{2\pi}{T_s}$, then we can write for $w_l = \frac{2\pi l}{T_s}$ radians/time, $l = 0, \pm 1, \pm 2, \ldots, \pm\frac{N-1}{2}$

$$
\begin{aligned}
G(jw_l) &= \lim_{\Delta t \to 0} \sum_{i=0}^{N-1} h(i\Delta t)\Delta t e^{-j\frac{2\pi il}{N}} \\
&= \lim_{\Delta t \to 0} \sum_{i=0}^{N-1} h_i e^{-j\frac{2\pi il}{N}}
\end{aligned}
\tag{4.9}
$$

where h_i are the discrete-time unit impulse response weights in Equation (4.1) and we have used the fact that $h_i \to h(i\Delta t)\Delta t$ as $\Delta t \to 0$. By comparing Equation (4.9) with the frequency response of the FIR model at $w = \frac{2\pi l}{N}$ radians using Equation (4.1), we are able to conclude that

$$
G(e^{j\frac{2\pi l}{N}}) \to G(jw_l)
\tag{4.10}
$$

as $\Delta t \to 0$.

Remark:

- The parameters of the FSF model correspond to the continuous-time process frequency response evaluated at $w = 0, \pm\frac{2\pi}{T_s}, \ldots, \pm\frac{\pi}{\Delta t}$ radians/time for a fixed value of T_s. Therefore, as Δt decreases, the number of parameters associated with the model increases, but only in the high frequency region.

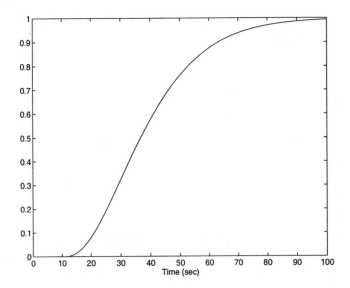

Figure 4.3: *Step response of the 3rd order plus delay system for Example 4.1*

Example 4.1. This example compares a discrete-time, rational transfer function model with the FSF model in terms of the effect of sampling rate on the model parameters. Consider a third order system with time delay given by

$$G(s) = \frac{e^{-10s}}{(10s + 1)^3} \qquad (4.11)$$

This system has a settling time of approximately 100 sec (i.e. $T_s = 100$), as can been seen from its unit step response shown in Figure 4.3. We choose three sampling intervals ($\Delta t = 0.5, 5, 10$ sec) for the comparison. Converting $G(s)$ in Equation (4.11) to a discrete-time, rational transfer function model, assuming a zero order hold on the input, gives:

$\Delta t = 0.5$ sec:

$$
\begin{aligned}
G(z) &= \frac{(0.2007 + 0.7732z^{-1} + 0.1862z^{-2}) \times 10^{-4}z^{-21}}{1.0000 - 2.8537z^{-1} + 2.7145z^{-2} - 0.8607z^{-3}} \\
&= \frac{0.2007 \times 10^{-4}(1 + 3.5949z^{-1})(1 + 0.2581z^{-1})z^{-21}}{(1 - 0.9512z^{-1})^3} \qquad (4.12)
\end{aligned}
$$

$\Delta t = 5$ sec:

$$
G(z) = \frac{(0.0144 + 0.0397z^{-1} + 0.0068z^{-2})z^{-3}}{1.0000 - 1.8196z^{-1} + 1.1036z^{-2} - 0.2231z^{-3}}
$$

$$= \frac{0.0144(1 + 2.5785z^{-1})(1 + 0.1831z^{-1})z^{-3}}{(1 - 0.6065z^{-1})^3} \qquad (4.13)$$

$\Delta t = 10$ sec:

$$G(z) = \frac{(0.0803 + 0.1544z^{-1} + 0.0179z^{-2})z^{-2}}{1.0000 - 1.1036z^{-1} + 0.4060z^{-2} - 0.0498z^{-3}}$$

$$= \frac{0.0803(1 + 1.7990z^{-1})(1 + 0.1238z^{-1})z^{-2}}{(1 - 0.3679z^{-1})^3} \qquad (4.14)$$

The parameters in the numerator of the discrete-time, rational transfer function model are small when the sampling interval is small but increase in magnitude as the sampling interval becomes larger. In the context of system identification, these smaller parameters are difficult to estimate accurately when noise is present in the data. Also, both numerator and denominator parameters in these transfer function models vary dramatically as the sampling interval changes and the numerator parameters, at least, do not have an obvious connection with the parameters of the underlying continuous-time system. In addition, all of the discrete-time transfer function models have two zeros with one located outside the unit circle, whereas the underlying continuous-time system has no zeros.

The FSF model parameters correspond to the frequency response of the discrete-time transfer function models presented in Equations (4.12)-(4.14) at three sets of frequencies:

(1) $w_1 = 0, \pm\frac{2\pi}{N_1}, \ldots, \pm\pi$ radians

(2) $w_2 = 0, \pm\frac{2\pi}{N_2}, \ldots, \pm\pi$ radians

(3) $w_3 = 0, \pm\frac{2\pi}{N_3}, \ldots, \pm\pi$ radians

where $N_1 = 201$, $N_2 = 21$ and $N_3 = 11$ corresponding to $N = \frac{T_s}{\Delta t}$ with N rounded up to the nearest odd number. Figure 4.4 shows the frequency response of the underlying continuous-time system and the FSF parameters for the three sampling intervals. These results illustrate that, as the sampling interval decreases, the parameters of the FSF model converge to the underlying continuous-time frequency response and, although the number of FSF parameters increases as the sampling interval becomes smaller, the additional parameters appear only in the high frequency region. Also, the variations in the FSF parameter values due to changes in the sampling interval are small compared to the changes observed with the parameters of the rational transfer function model.

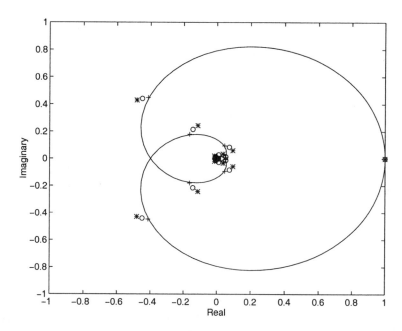

Figure 4.4: *Continuous-time frequency response (solid) and FSF model parameters for Example 4.1 ('+': $\Delta t = 0.5$; 'o': $\Delta t = 5$; '*': $\Delta t = 10$)*

4.4 REDUCED ORDER FSF MODEL

The number of parameters in the FSF model in Equation (4.4) is equal to N, the same number associated with the impulse response model in Equation (4.1). However, in Example 4.1 we see that the majority of the FSF parameters lie near the origin of the complex plane, i.e. they have small magnitudes. This is a result of the fact that the continuous-time frequency response for this example converges to zero at high frequencies. The class of processes that have negligible frequency response magnitudes in the higher frequency region are those processes with transfer functions having a numerator order less than the denominator order, referred to as strictly proper transfer functions. Given that we have already assumed the process to be stable and if we further assume that its transfer function is strictly proper, then the continuous-time frequency response of the process $|G(jw)| \rightarrow 0$ as $w \rightarrow \infty$. Thus, based on the relationship between the FSF parameters and the continuous-time frequency response stated by Theorem 4.1, we can conclude that as $\Delta t \rightarrow 0$:

- there exists an odd integer n such that for all l where $\frac{n-1}{2} < |l| \leq \frac{N-1}{2}$, the magnitudes of the FSF model parameters, $|G(e^{j\frac{2\pi l}{N}})|$, are approximately equal to zero;

- n becomes independent of the choice of the sampling interval.

Based on these properties, we will refer to n as the reduced order of the FSF model, which represents the number of significant parameters in the FSF model. We must qualify the use of the term "reduced order" to distinguish the number of significant parameters in the FSF model (n) from the order of the individual FSF filters (N). This reduced nth order FSF model can be written in the following form

$$G(z) \approx \sum_{l=-\frac{n-1}{2}}^{\frac{n-1}{2}} G(e^{j\frac{2\pi l}{N}}) \frac{1}{N} \frac{1 - z^{-N}}{1 - e^{j\frac{2\pi l}{N}} z^{-1}} \qquad (4.15)$$

Due to the structure of the FSF model, the terms in Equation (4.4) that have been neglected in the reduced order model of Equation (4.15) (i.e. $G(e^{j\frac{2\pi l}{N}})$ for $\frac{n-1}{2} < |l| \leq \frac{N-1}{2}$) always correspond to the high frequency dynamics of the process. For instance, in Example 4.1, with a choice of sampling interval $\Delta t = 0.5$ ($N = 201$), there are at most 9 FSF parameters ($n = 9$) that can be distinguished from the origin of the complex plane (see Figure 4.4). The remaining 192 parameters ($201 - 9 = 192$) in the high frequency region are very close to zero and can perhaps be neglected.

Intuitively, it would seem that the reduced model order n is related to the properties of the underlying continuous-time system. For instance, it would appear that its value depends on how fast the magnitude of the process frequency response rolls off to zero. We will illustrate this point in the following example.

Example 4.2. Consider the process transfer function

$$G(s) = \frac{e^{-ds}}{(Ts + 1)^q} \qquad (4.16)$$

where $q = 1, 2, 3$ and $d \leq T$. The settling times T_s for these processes, including the effect of the time delay, are approximated as $6T$, $8T$ and $10T$ corresponding to $q = 1$, 2 and 3, respectively. The frequency responses of these processes at $w_l = \frac{2\pi l}{T_s}$ radians/time are given by, for the first order system

$$G(jw_l) = \frac{e^{-jd\frac{\pi}{3T}l}}{j\frac{\pi}{3}l + 1} \qquad (4.17)$$

q	T_s	n (10%)	n (5%)	n (1%)
1	$6T$	19	39	191
2	$8T$	7	11	25
3	$10T$	7	9	15

Table 4.1: *Reduced FSF model orders corresponding to different truncation levels*

for the second order system

$$G(jw_l) = \frac{e^{-jd\frac{\pi}{4T}l}}{(j\frac{\pi}{4}l + 1)^2} \tag{4.18}$$

and for the third order system

$$G(jw_l) = \frac{e^{-jd\frac{\pi}{5T}l}}{(j\frac{\pi}{5}l + 1)^3} \tag{4.19}$$

The reduced FSF model order n corresponding to maximum truncation levels of 10%, 5% and 1% are listed in Table 4.1 for the above examples, where the neglected frequency response coefficients all have magnitudes less than the indicated percentage of the steady state gain. Note that the reduced orders given in Table 4.1 are independent of the time constant T.

Effect of FSF Model Reduction on Process Frequency Response

We can attempt to construct the process frequency response using the reduced order FSF model given in Equation (4.15) by letting $z = e^{jw}$ and evaluating

$$G(e^{jw}) = \sum_{l=-\frac{n-1}{2}}^{\frac{n-1}{2}} G(e^{j\frac{2\pi l}{N}})\frac{1}{N}\frac{1 - e^{-jwN}}{1 - e^{j\frac{2\pi l}{N}}e^{-jw}} \tag{4.20}$$

From the properties of the frequency sampling filters, there will be no errors at the respective centre frequencies of the filters, i.e. $G(e^{jw}) = G(e^{j\frac{2\pi l}{N}})$, for $l = 0, \pm1, \pm2, \ldots, \pm\frac{n-1}{2}$. However, there will be errors at all frequencies other than the centre frequencies. These errors can be expressed in terms of the following equation with $z = e^{jw}$

$$\Delta G(z) = \sum_{l=-\frac{N-1}{2}}^{-\frac{n+1}{2}} G(e^{j\frac{2\pi l}{N}})\frac{1}{N}\frac{1 - z^{-N}}{1 - e^{j\frac{2\pi l}{N}}z^{-1}} + \sum_{l=\frac{n+1}{2}}^{\frac{N-1}{2}} G(e^{j\frac{2\pi l}{N}})\frac{1}{N}\frac{1 - z^{-N}}{1 - e^{j\frac{2\pi l}{N}}z^{-1}}$$

$$\tag{4.21}$$

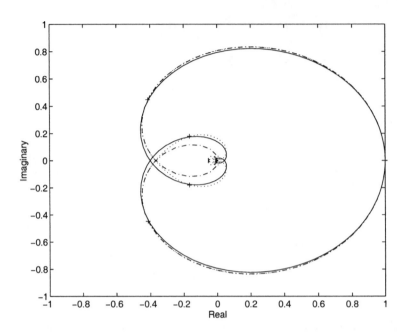

Figure 4.5: *Construction of frequency responses using reduced order FSF models for Example 4.3 (solid: true response; dash-dotted: $n = 3$; dotted: $n = 5$)*

which represents a summation of the neglected FSF parameters weighted by the frequency responses of their respective filters. Given that the magnitude of the frequency response of each filter decays as the frequencies move away from the filter's centre frequency (see Figure 4.2), the size of the error at a particular frequency depends on the distance from this frequency to the neglected high frequency FSF parameters and the magnitudes of these neglected parameters. Therefore, we would expect that, in the lower frequency region where the frequencies are far away from the high frequency region containing the neglected parameters, the error should be small.

Example 4.3. Let us consider again the process given by Equation (4.11). We will choose $\Delta t = 0.5$ and $N = 201$ for this example. Figure 4.5 shows the construction of the process frequency response using two different reduced order FSF models in Equation (4.20). For $n = 3$, there are large errors in the constructed frequency response in the higher frequency region. However, with $n = 5$, the accuracy of the construction in the higher frequency region improves significantly. Also, it appears that for both $n = 3$ and 5, the truncated high frequency terms have similar but relatively little negative impact on the accuracy in the frequency region between $w = 0$ and $\pm\frac{2\pi}{N}$.

4.5   PARAMETER ESTIMATION FOR THE FSF MODEL

This section deals with the problem of estimating the parameters of a reduced order FSF model from process input-output data using the least
squares algorithm.

Single-Input, Single-Output Systems

We assume that the process being identified is stable, linear and time invariant and can be accurately represented by a reduced nth order FSF model.
For an arbitrary process input $u(k)$ and the measured process output $y(k)$,
the frequency sampling filter model can be written as

$$y(k) = G(z)u(k) + \xi(k) \qquad (4.22)$$

where $G(z)$ is given by Equation (4.15) and $\xi(k)$ is a disturbance term. The
process output can be expressed in a linear regression form by defining the
parameter vector as

$$\theta = [G(e^{j0}) \;\; G(e^{j\frac{2\pi}{N}}) \;\; G(e^{-j\frac{2\pi}{N}}) \;\; \cdots \;\; G(e^{j\frac{(n-1)\pi}{N}}) \;\; G(e^{-j\frac{(n-1)\pi}{N}})]^T$$

and the regressor vector as

$$\phi(k) = [f(k)^0 \;\; f(k)^1 \;\; f(k)^{-1} \;\; \cdots \;\; f(k)^{\frac{n-1}{2}} \;\; f(k)^{-\frac{n-1}{2}}]^T$$

where

$$f(k)^r = \frac{1}{N} \frac{1 - z^{-N}}{1 - e^{j\frac{2\pi r}{N}} z^{-1}} u(k) \qquad (4.23)$$

for $r = 0, \pm 1, \ldots, \pm\frac{n-1}{2}$. This allows us to rewrite Equation (4.22) as

$$y(k) = \phi(k)^T \theta + \xi(k) \qquad (4.24)$$

Note that:

- the elements of the parameter vector θ are arranged in such a manner that the parameter associated with the zero frequency enters first,
  followed by the first complex conjugate pair of frequency parameters,
  followed by the second complex conjugate pair of frequency parameters, and so on;

- the elements of the regressor vector are generated by passing the
  process input through the corresponding frequency sampling filters.

In order to formulate the least squares solution, Equation (4.24) is written in a matrix form for M pairs of input-output data

$$Y = \Phi\theta + \zeta \tag{4.25}$$

where $Y^T = [y(0) \; y(1) \; \cdots \; y(M-1)]$, $\zeta^T = [\xi(0) \; \xi(1) \; \cdots \; \xi(M-1)]$ and

$$\Phi = \begin{bmatrix} f(0)^0 & f(0)^1 & f(0)^{-1} & \cdots & f(0)^{-\frac{n-1}{2}} \\ f(1)^0 & f(1)^1 & f(1)^{-1} & \cdots & f(1)^{-\frac{n-1}{2}} \\ \vdots & \vdots & \vdots & & \vdots \\ f(M-1)^0 & f(M-1)^1 & f(M-1)^{-1} & \cdots & f(M-1)^{-\frac{n-1}{2}} \end{bmatrix}$$

The least squares estimates of the FSF model parameters are given by

$$\hat{\theta} = (\Phi^*\Phi)^{-1}\Phi^*Y \tag{4.26}$$

which minimizes the sum of squared prediction errors

$$V = (Y - \Phi\theta)^T(Y - \Phi\theta) \tag{4.27}$$

where (*) denotes the complex conjugate transpose.

Multi-Input, Multi-Output Systems

Although we could extend the least squares estimation results for SISO systems directly to the p-input, q-output multivariable case, we prefer to treat these systems as q multi-input, single-output (MISO) systems. This way, we can take full advantage of the orthogonal decomposition algorithm developed in Chapter 3 for parameter estimation and structure selection of the p subsystems associated with each of the q outputs. This will be illustrated using an industrial data set in Chapter 5.

For each process output, the p inputs are denoted as $u_1(k)$, $u_2(k),\ldots,$ $u_p(k)$, the times to steady state for the individual subsystems are given by $N_1, N_2,\ldots, N_p$, and the reduced orders for each subsystem represented by its own FSF model are chosen to be $n_1, n_2, \ldots, n_p$. In the matrix representation for this MISO system, the first input $u_1(k)$ is passed through a set of n_1 frequency sampling filters based on N_1 to form the first n_1 columns in the data matrix Φ, followed by passing the second input $u_2(k)$ through a set of n_2 frequency sampling filters based on N_2 to form the next n_2 columns in the data matrix Φ, etc. The parameter vector θ contains the n_1 frequency response parameters associated with the first subsystem, followed by the n_2

frequency response parameters associated with the second subsystem, etc. Using this matrix representation, the least squares algorithm can be directly applied to estimate the FSF model parameters associated with each of the p subsystems.

4.6 NATURE OF THE CORRELATION MATRIX

The conditioning of the data matrix Φ in Equation (4.25) provides information on the potential difficulties to be encountered in calculations based on Φ. (The condition number for any matrix X is defined as $\frac{\mu_{max}}{\mu_{min}}$, where μ_{min} and μ_{max} denote the minimum and maximum singular values of X, respectively.) For instance, the worse the conditioning of the Φ matrix, as indicated by a large condition number relative to unity, the greater the potential that small relative changes in the output data Y will result in large relative changes in the least squares parameter estimates generated in Equation (4.26) (Belsley, 1991). This follows from the fact that the covariance matrix associated with the least squares estimates is $(\Phi^*\Phi)^{-1}\sigma^2$, where σ^2 is the variance of ζ. Instead of performing an examination of the condition number of the data matrix Φ, we choose to directly examine the relative values of the diagonal and off-diagonal elements in the matrix $\Phi^*\Phi$, referred to as the correlation matrix. The objective here is to see if this will provide insight into whether the correlation matrix is ill-conditioned, resulting in blow-up of the covariance matrix and, in turn, large uncertainty in the parameter estimates.

The elements of the correlation matrix in Equation (4.26) are given by $\sum_{k=0}^{M-1} f(k)^{r*} f(k)^q$ where r and q are integers in the range $[-\frac{n-1}{2}, +\frac{n-1}{2}]$. The diagonal elements correspond to $r = q$ and the off-diagonal elements correspond to $r \neq q$. We show in the following theorem that the elements of the correlation matrix are weighted sums of the energy contributions from different frequencies present in the input signal.

Theorem 4.2: Let $U(e^{jw'_l})$ denote the DFT of the input signal $u(k)$ with $w'_l = \frac{2\pi l}{M}$ radians. Let $H^r(e^{jw'_l})$ denote the frequency response of the rth filter given by

$$H^r(e^{jw'_l}) = \frac{1}{N} \frac{1 - e^{-jw'_l N}}{1 - e^{j\frac{2\pi r}{N}} e^{-jw'_l}} \tag{4.28}$$

Then, the diagonal elements of the correlation matrix are

$$\sum_{k=0}^{M-1} f(k)^{r*} f(k)^r = \sum_{l=0}^{M-1} \frac{1}{N^2} \frac{sin^2(\frac{w'_l N}{2})}{sin^2(\frac{w'_l - \frac{2\pi r}{N}}{2})} |U(e^{jw'_l})|^2 \tag{4.29}$$

and the off-diagonal elements are, for $r \neq q$

$$\sum_{k=0}^{M-1} f(k)^{r*} f(k)^q = \sum_{l=0}^{M-1} H^r(e^{jw_l'})^* H^q(e^{jw_l'}) |U(e^{jw_l'})|^2 \qquad (4.30)$$

with

$$\left| \sum_{k=0}^{M-1} f(k)^{r*} f(k)^q \right| \leq \sum_{l=0}^{M-1} \frac{1}{N^2} \frac{sin^2\left(\frac{w_l'N}{2}\right)}{\left| sin\left(\frac{w_l' - \frac{2\pi r}{N}}{2}\right) \right| \left| sin\left(\frac{w_l' - \frac{2\pi q}{N}}{2}\right) \right|} |U(e^{jw_l'})|^2$$

$$(4.31)$$

Proof: From Parseval's theorem, we can write the following relationship for the diagonal elements of the correlation matrix

$$\sum_{k=0}^{M-1} f(k)^{r*} f(k)^r = \sum_{l=0}^{M-1} |H^r(e^{jw_l'})|^2 |U(e^{jw_l'})|^2 \qquad (4.32)$$

Then, Equation (4.29) follows from

$$|H^r(e^{jw_l'})|^2 = \frac{1}{N^2} \left| \frac{1 - e^{-jw_l'N}}{1 - e^{j\frac{2\pi r}{N}} e^{-jw_l'}} \right|^2 \qquad (4.33)$$

$$= \frac{1}{N^2} \frac{1 - cos(w_l'N)}{1 - cos(w_l' - \frac{2\pi r}{N})} \qquad (4.34)$$

$$= \frac{1}{N^2} \frac{sin^2\left(\frac{w_l'N}{2}\right)}{sin^2\left(\frac{w_l' - \frac{2\pi r}{N}}{2}\right)} \qquad (4.35)$$

where we have used the identity $1 - cos\alpha = 2sin^2\left(\frac{\alpha}{2}\right)$.

For the off-diagonal elements $(r \neq q)$, similar analysis can be applied to obtain Equations (4.30) and (4.31).

Remark:

- The value $|U(e^{jw_l'})|^2$ is known as the periodogram of the input signal $u(k)$ and is a measure of the energy contribution of the frequency w_l' to the input signal (Ljung, 1987). Therefore, these results relate the elements in the correlation matrix directly to the periodogram of the input signal through a weighted sum relationship.

Weighting Function Associated with Diagonal Elements

Let

$$W^r(w_l') = \frac{1}{N^2} \frac{sin^2\left(\frac{w_l' N}{2}\right)}{sin^2\left(\frac{w_l' - \frac{2\pi r}{N}}{2}\right)} \tag{4.36}$$

denote the weighting function in Equation (4.29) associated with a particular value of r. The properties of these weighting functions determine the contributions that the input signal energy makes to each diagonal element.

Property A. Magnitude at the centre frequency
The magnitude of the weighting function at the centre frequency of the filter is equal to unity, i.e.

$$\lim_{w_l' \to \frac{2\pi r}{N}} W^r = 1 \tag{4.37}$$

Property B. Zeros of the weighting function
At $w_l' = \frac{2\pi r}{N} \pm \frac{2\pi m}{N}$, $W^r(w_l') = 0$, where m is a positive integer. This means that the weighting function is zero at the centre frequencies of the other filters. The distance between the two zeros of $W^r(w_l')$ closest to the centre frequency $(w_l' = \frac{2\pi r}{N} \pm \frac{2\pi}{N})$ forms the bandwidth of the weighting function and is equal to $\frac{4\pi}{N}$.

Proof of Property A: At $w_l' = \frac{2\pi r}{N}$

$$sin^2\left(\frac{w_l' N}{2}\right) = 0 \tag{4.38}$$

and

$$sin^2\left(\frac{w_l' - \frac{2\pi r}{N}}{2}\right) = 0 \tag{4.39}$$

Therefore, W^r becomes undefined at this frequency. Instead, we will return to the original expression for the weighting function given by Equation (4.32) and use the series expansion of $H^r(e^{jw_l'})$ to obtain

$$H^r(e^{jw_l'}) = \frac{1}{N} \sum_{i=0}^{N-1} e^{j\frac{2\pi r i}{N}} e^{-jw_l' i} \tag{4.40}$$

where it can now be seen that

$$\lim_{w_l' \to \frac{2\pi r}{N}} H^r(e^{jw_l'}) = 1 \tag{4.41}$$

Hence Property A is true.

Proof of Property B: At $w_l' = \frac{2\pi r}{N} \pm \frac{2\pi m}{N}$

$$sin^2\left(\frac{w_l' N}{2}\right) = sin^2(\pi r \pm \pi m) = 0 \qquad (4.42)$$

and

$$sin^2\left(\frac{w_l' - \frac{2\pi r}{N}}{2}\right) = sin^2\left(\pm\frac{\pi m}{N}\right) \neq 0 \qquad (4.43)$$

Hence, Property B is true.

Remark:

- Figure 4.6 shows one of the weighting functions W^r with $r = 4$. For a different value of r, the shape of the weighting function remains the same, but the location of the centre frequency changes to $\frac{2\pi r}{N}$ radians. It can be seen that W^r has a very narrow bandpass nature and that it behaves like a delta function at the centre frequency. As a result, this weighting function yields a corresponding diagonal element in the correlation matrix that is largely determined by the energy content of the input signal in the vicinity of the frequency $\frac{2\pi r}{N}$.

With a modest assumption on the periodogram of the input signal, we can obtain a simplified expression for the diagonal elements of the correlation matrix.

Theorem 4.3: If we assume that the periodogram of the input signal is approximately equal to a constant value $(|U(e^{jw_i'})|^2 \approx \bar{U})$ over the narrow frequency region from $\frac{2\pi(r-1)}{N}$ to $\frac{2\pi(r+1)}{N}$, then

$$\sum_{k=0}^{M-1} f(k)^{r*} f(k)^r \approx \frac{M}{N}\bar{U} \qquad (4.44)$$

Proof: Using Properties A and B, we know that W^r achieves its maximum value of unity at the frequency $\frac{2\pi r}{N}$ and is equal to zero at $\frac{2\pi r}{N} \pm \frac{2\pi}{N}$. We assume that the magnitude of W^r outside this frequency interval is very small. Thus, the most significant contribution to the diagonal element comes from the frequency content of the input signal within the frequency interval

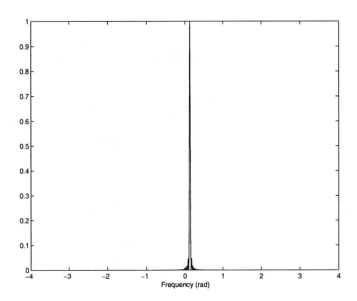

Figure 4.6: *Weighting function associated with a diagonal element of the FSF correlation matrix (r = 4 and N = 201)*

$\frac{2\pi r}{N} - \frac{2\pi}{N} \leq w_l' \leq \frac{2\pi r}{N} + \frac{2\pi}{N}$. Therefore,

$$\sum_{k=0}^{M-1} f(k)^{r*} f(k)^r = \sum_{l=0}^{M-1} |H^r(e^{jw_l'})|^2 |U(e^{jw_l'})|^2$$

$$\approx \sum_{l=\frac{M}{N}(r-1)}^{\frac{M}{N}(r+1)} |H^r(e^{jw_l'})|^2 |U(e^{jw_l'})|^2 \qquad (4.45)$$

where we have restricted the frequency range used in the summation to $w_l' = \frac{2\pi l}{M} = \frac{2\pi r}{N} \pm \frac{2\pi}{N}$. With the assumption that the periodogram of the input signal is constant over this frequency region, Equation (4.45) simplifies to

$$\sum_{k=0}^{M-1} f(k)^{r*} f(k)^r \approx \bar{U} \left(\sum_{l=\frac{M}{N}(r-1)}^{\frac{M}{N}(r+1)} |H^r(e^{jw_l'})|^2 \right) \qquad (4.46)$$

We now choose to approximate the weighting function by a symmetric triangle with height equal to unity and width equal to $\frac{2M}{N}$. Then, the summation on the right-hand side of Equation (4.46) becomes equivalent to the area of this triangle ($\frac{M}{N}$). This concludes the proof.

Remarks:

- The diagonal elements of the correlation matrix are proportional to the data length M.

- The diagonal element corresponding to a particular value of r is proportional to the periodogram of the input signal in the vicinity of the frequency $\frac{2\pi r}{N}$. Therefore, if the input signal does not contain significant energy in the vicinity of the centre frequency for the rth filter, then the corresponding diagonal element will be small.

Weighting Function Associated with Off-Diagonal Elements

Let the weighting function in the upper bound on the magnitude of the off-diagonal elements in Equation (4.31) be denoted by

$$W^{r,q}(w_l') = \frac{1}{N^2} \frac{sin^2\left(\frac{w_l'N}{2}\right)}{\left|sin\left(\frac{w_l' - \frac{2\pi r}{N}}{2}\right)\right|\left|sin\left(\frac{w_l' - \frac{2\pi q}{N}}{2}\right)\right|} \tag{4.47}$$

Figure 4.7 illustrates the weighting functions $W^{r,q}$ with $r = 4$ and $q = 5$, 6, 7. The shape of $W^{r,q}$ somewhat resembles that of W^r, with two exceptions. It is actually bimodal in shape and its maximum value is less than unity. For $|r - q| = 1$, 2, 3 with $r = 4$, the maximum values of $W^{r,q}$ are approximately 0.4, 0.14, 0.09 and hence the maximum value decreases as $|r - q|$ increases. Since the magnitude of the off-diagonal elements are bounded by the periodogram of the input signal weighted by $W^{r,q}$, their values will decrease rapidly as $|r - q|$ increases, relative to the diagonal element W^r. In general, the magnitudes of the off-diagonal elements are smaller than the corresponding diagonal element, in both the row and column directions, and their magnitudes decrease as their distance from the diagonal increases.

Remarks:

- With the analysis of the diagonal and off-diagonal elements in the correlation matrix obtained when using a frequency sampling filter model structure, we can conclude that the magnitudes of the diagonal elements approximately reflect the conditioning of the correlation matrix. In other words, if the ratio between the largest and smallest diagonal elements is large, then the correlation matrix is likely ill-conditioned. Therefore, an obvious way to attempt to improve the conditioning of the correlation matrix is to eliminate the frequency sampling filters

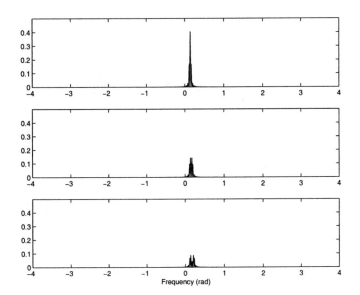

Figure 4.7: *Weighting functions associated with off-diagonal elements of the FSF correlation matrix (r = 4 and N = 201: q = 5 (top); q = 6 (middle); q = 7 (bottom))*

that result in small diagonal elements relative to the large diagonal elements.

- To increase the magnitude of a diagonal element corresponding to a particular value of r, the periodogram of the input signal has to be increased in the vicinity of the centre frequency of the rth frequency sampling filter.

Example 4.4. We will illustrate the properties of the diagonal and off-diagonal elements in the correlation matrix by comparing the FSF result with the correlation matrix that would result with an FIR model structure using an identical input signal. Consider the process described by the transfer function

$$G(s) = \frac{e^{-20s}}{(20s + 1)^2} \tag{4.48}$$

This process has an approximate settling time of 199 sec and is sampled with an interval of 1 sec. Thus the parameter N is chosen to be 199 for both the FSF and FIR models. For the identification experiment, we have used a binary input signal with amplitude equal to unity. The input signal has been taken as a generalized random binary signal (GRBS) with probability

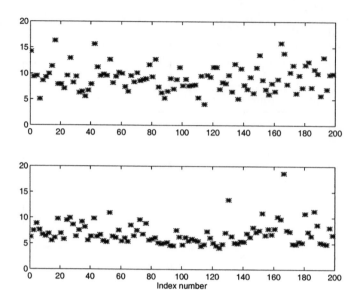

Figure 4.8: *Elements of the correlation matrix using an FSF model structure (dimension* 199×199*) for Example 4.4. Upper diagram: the diagonal elements of the correlation matrix; lower diagram: row sums of the absolute values of the off-diagonal elements*

of switching equal to 0.5 (Tulleken, 1990). This produces an input signal with frequency response characteristics close to white noise.

This sequence was passed through all N frequency sampling filters to produce the correlation matrix for the FSF case with $n = N$. Figure 4.8 shows the diagonal elements and the row sums of the absolute values of the off-diagonal elements. Given that the periodogram for this type of input signal is relatively constant over the entire frequency range, the diagonal elements of the correlation matrix are approximately equal. Strictly speaking, this correlation matrix is not diagonally dominant because the summed values of the off-diagonal elements shown in the lower diagram are not all less than the corresponding diagonal elements in the upper diagram. However, we did find that the magnitudes of the individual off-diagonal elements are all smaller than the corresponding diagonal element. With these characteristics, we would expect the correlation matrix to be well-conditioned. In fact, this matrix is well-conditioned with a condition number of 4.72.

The same input signal was passed through a set of simple delay elements up to $z^{-(N-1)}$ to produce the correlation matrix for the FIR case. We evaluated diagonal and off-diagonal elements of this correlation matrix and the

results are shown in Figure 4.9. The conclusion is that the individual elements of the correlation matrix in the FIR case do not provide us with
any information concerning its numerical conditioning. This is evident from
Figure 4.9 where it is seen that the diagonal elements are identical and the
sums of the absolute values of the off-diagonal elements are all much greater
than the diagonal elements. However, the condition number of the correlation matrix (4.72) is identical to that for the FSF case, as expected, because
the full order FSF model is only a linear transformation of the FIR model.

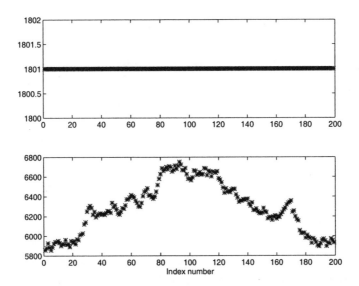

Figure 4.9: *Elements of the correlation matrix using an FIR model structure (dimension 199 × 199) for Example 4.4. Upper diagram: the diagonal elements of
the correlation matrix; lower diagram: row sums of the absolute values of the off-
diagonal elements*

Chapter 5

From FSF Models to Step Response Models

5.1 INTRODUCTION

The previous chapter introduced the frequency sampling filter model structure in the context of process identification. This chapter pursues the topic further by focusing on the identification of step response models using the FSF model structure.

This chapter consists of seven sections. Section 5.2 shows how to obtain an estimate of the process step response from the FSF model parameters. Section 5.3 discusses the topic of smoothing the step response estimate using a reduced order FSF model. Section 5.4 analyzes the errors introduced by using a reduced order model. Section 5.5 derives confidence bounds for the process frequency response and step response estimates obtained from an FSF model. Section 5.6 presents a generalized least squares algorithm for identification of multi-input, single-output systems in which the *PRESS* criterion is used to determine both the FSF process model order for each subsystem and the number of terms to be included in the noise model. Section 5.7 presents an industrial case study using data collected from the Sunoco refinery in Sarnia, Canada.

Portions of this chapter have been reprinted from *Automatica* **33**, L. Wang and W.R. Cluett, "Frequency-sampling filters: an improved model structure for step-response identification", pp. 939-944, 1997, with permission from Elsevier Science, and from *Journal of Process Control* **7**, W.R. Cluett, L. Wang and A. Zivkovic, "Development of quality bounds for time and frequency domain models: application to the Shell distillation column", pp. 75-80, 1997, with permission from Elsevier Science.

5.2   OBTAINING A STEP RESPONSE MODEL FROM THE FSF MODEL

Let g_m denote the unit step response of the process at sampling instant m, $m = 0, \ldots, N - 1$. Then, g_m is related to the discrete-time unit impulse response coefficients h_i, $i = 0, \ldots, N - 1$ according to

$$g_m = \sum_{i=0}^{m} h_i \tag{5.1}$$

From Equation (4.2)

$$h_i = \frac{1}{N} \sum_{l=-\frac{N-1}{2}}^{\frac{N-1}{2}} G(e^{j\frac{2\pi l}{N}}) e^{j\frac{2\pi l i}{N}} \tag{5.2}$$

we can write

$$g_m = \frac{1}{N} \sum_{i=0}^{m} \sum_{l=-\frac{N-1}{2}}^{\frac{N-1}{2}} G(e^{j\frac{2\pi l}{N}}) e^{j\frac{2\pi l i}{N}} \tag{5.3}$$

Given that $G(e^{j\frac{2\pi l}{N}})$ is independent of the index i, we can interchange the summations and rewrite Equation (5.3) as

$$g_m = \sum_{l=-\frac{N-1}{2}}^{\frac{N-1}{2}} G(e^{j\frac{2\pi l}{N}}) \frac{1}{N} \frac{1 - e^{j\frac{2\pi l}{N}(m+1)}}{1 - e^{j\frac{2\pi l}{N}}} \tag{5.4}$$

where we have used the relation

$$\sum_{i=0}^{m} e^{j\frac{2\pi l i}{N}} = \frac{1 - e^{j\frac{2\pi l}{N}(m+1)}}{1 - e^{j\frac{2\pi l}{N}}} \tag{5.5}$$

Equation (5.4) gives an explicit relationship between the parameters of the FSF model and the step response coefficients. The step response coefficients, evaluated using a reduced order FSF model, can be obtained from

$$g_m \approx \sum_{l=-\frac{n-1}{2}}^{\frac{n-1}{2}} G(e^{j\frac{2\pi l}{N}}) \frac{1}{N} \frac{1 - e^{j\frac{2\pi l}{N}(m+1)}}{1 - e^{j\frac{2\pi l}{N}}} \tag{5.6}$$

where we have simply replaced N in Equation (5.4) by n, and thereby have neglected the high frequency parameters of the FSF model.

Now the question arises as to how neglecting these high frequency parameters affects the accuracy of the step response coefficients. To study this problem, we define

$$S(l,m) = \frac{1}{N} \frac{1 - e^{j\frac{2\pi l}{N}(m+1)}}{1 - e^{j\frac{2\pi l}{N}}} \tag{5.7}$$

that, for $l \neq 0$, has a real part

$$S_R(l,m) = \frac{1}{2N(1 - cos(w_l))} \left(1 - cos(w_l) + 2sin(\frac{w_l}{2})sin\left(\frac{2\pi lm + \pi l}{N}\right)\right) \tag{5.8}$$

and an imaginary part

$$S_I(l,m) = \frac{1}{2N(1 - cos(w_l))} \left(sin(w_l) - 2sin(\frac{w_l}{2})cos\left(\frac{2\pi lm + \pi l}{N}\right)\right) \tag{5.9}$$

where $w_l = \frac{2\pi l}{N}$. Therefore, Equation (5.6) can be rewritten in terms of its real and imaginary parts as

$$g_m = \frac{1}{N} \sum_{l=0}^{\frac{n-1}{2}} [Real(G(e^{j\frac{2\pi l}{N}}))S_R(l,m) - Imag(G(e^{j\frac{2\pi l}{N}}))S_I(l,m)] \tag{5.10}$$

Hence, $S_R(l,m)$ and $S_I(l,m)$ act as weighting functions on the respective real and imaginary parts of the process frequency response when generating the step response coefficients. Figure 5.1 shows the behaviour of the weighting functions $S_R(l,m)$ for $N = 200$ and $l = 0,1,2,7$. At $l = 0$, this weighting function starts at $\frac{1}{N}$ for $m = 0$ and increases in a linear fashion until it reaches a value of 1.0 at $m = 199$. Figure 5.2 shows the behaviour of $S_I(l,m)$ for $N = 200$ and $l = 1,2,7$. For $l = 0$, $S_I(l,m)$ is equal to zero for all m. For $l > 0$, both $S_R(l,m)$ and $S_I(l,m)$ have their largest magnitudes when $l = 1$ and hence $G(e^{j\frac{2\pi}{N}})$ contributes most to the step response beyond $G(e^{j0})$. The next most significant term corresponds to $l = 2$. The weighting functions for $l = 7$ indicate that as l increases, the contributions from the corresponding higher frequencies parameters of the FSF model to the step response coefficients decrease.

Two examples are now given to demonstrate the construction of step response models from the FSF model structure.

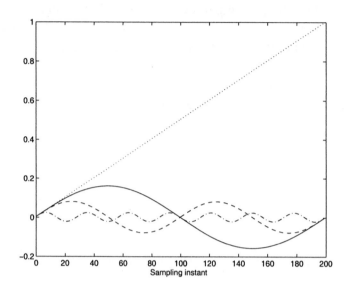

Figure 5.1: *Real part of weighting functions relating frequency response to step response (dotted: $l = 0$; solid: $l = 1$; dashed: $l = 2$; dash-dotted: $l = 7$)*

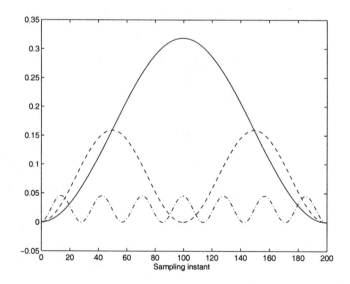

Figure 5.2: *Imaginary part of weighting function relating frequency response to step response (solid: $l = 1$; dashed: $l = 2$; dash-dotted: $l = 7$)*

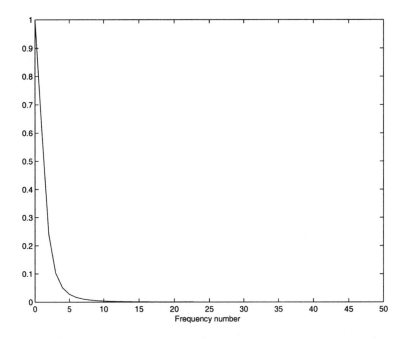

Figure 5.3: *Magnitudes of the FSF model parameters for Example 5.1*

Example 5.1. Consider the process given by the transfer function

$$G(s) = \frac{e^{-10s}}{(10s + 1)^3} \tag{5.11}$$

Using a sampling interval $\Delta t = 0.5$, we have calculated the parameters of the corresponding FSF model using a settling time estimate of 100 sec. Figure 5.3 shows that the magnitudes of the FSF model parameters decay very quickly for this process. We have constructed the process step response using Equation (5.6) for $n = 3, 5, 7, 9$. The responses with $n = 3$ and 5 are compared to the true step response in Figure 5.4. For the different choices of n, the sum of squared errors between the true step response coefficients and the ones generated from the reduced order FSF models are: $0.5271(n = 3)$, $0.0508(n = 5)$, $0.0056(n = 7)$ and $0.0032(n = 9)$. The responses with $n = 7$ and 9 have not been shown because they lie almost exactly on the true response. It is interesting to note that even a value of $n = 7$, which corresponds to a seemingly large truncation error of 10% in the frequency domain (see Table 4.1), produces an accurate representation of the step response.

In comparison with the FIR model, the FSF model is able to represent

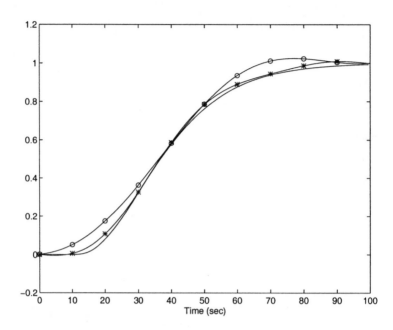

Figure 5.4: *Comparison of true step response with those generated from reduced order FSF models for Example 5.1 (solid: true response; 'o': $n = 3$; '*': $n = 5$)*

the step response with many fewer parameters. For instance, even with a larger sampling interval of 2 sec, the FIR model would require at least 50 parameters to capture the process dynamics with a similar degree of accuracy.

Example 5.2. Consider the process given by the transfer function

$$G(s) = \frac{K_p R(\tau_2 s + 1)e^{-d_1 s}}{(\tau_1 s + 1)(\tau_2 s + 1 - (1 - R)e^{-d_2 s})} \tag{5.12}$$

where $d_1 = 45$, $d_2 = 75$, $K_p = 0.8$, $R = 0.6$, $\tau_1 = 5$ and $\tau_2 = 10$. Dynamics such as these arise with processes that have recycle streams. Choosing $\Delta t = 1$, we have calculated the parameters of the corresponding FSF model using a settling time estimate of 400 sec. Figure 5.5 shows how the amplitudes of the FSF model parameters decay very slowly for this process. Using $n = 99$ parameters (equivalent to 50 frequencies), we have constructed the process step response using Equation (5.6) and the result is compared to the true step response in Figure 5.6. The sum of squared errors between the true step response coefficients and the ones generated from the FSF model with $n = 99$ is 0.0023.

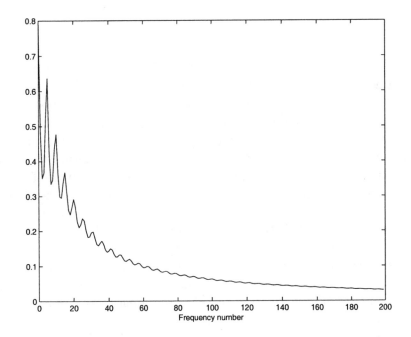

Figure 5.5: *Magnitudes of the FSF model parameters for Example 5.2*

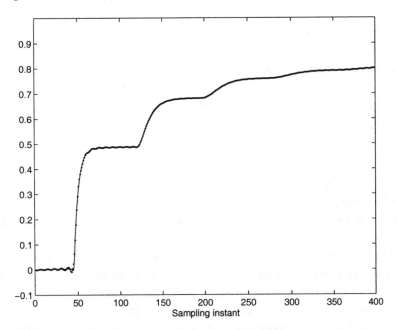

Figure 5.6: *Comparison of true step response with that generated from a reduced order FSF model for Example 5.2 (solid: true response; dotted: $n = 99$)*

Remark:

- These two examples demonstrate that the number of parameters required by the FSF model to accurately construct a step response model is determined by the underlying continuous-time process frequency response. If the process frequency response is relatively smooth and decays quickly to zero at higher frequencies, such as the process in Example 5.1, then the FSF model structure can capture the process dynamics with a very small number of parameters. If the frequency response is complicated and decays slowly to zero at higher frequencies, such as the process in Example 5.2, the number of parameters required in the FSF structure increases. Nevertheless, the FSF structure provides an effective means to describe various types of process dynamics without the need for process structural information and is, in many cases, more efficient than the FIR model structure in terms of the number of parameters required to represent a process with a given accuracy.

5.3   SMOOTHING THE STEP RESPONSE USING THE FSF MODEL

This section is devoted to a simulation example that illustrates the problems associated with obtaining an estimate of the process step response using an FIR model and motivates the use of a reduced order FSF model instead.

Example 5.3. Consider the process given by the transfer function

$$G(s) = \frac{e^{-20s}}{(20s + 1)^2} \tag{5.13}$$

This process is sampled with an interval of 2 sec, and the number of samples required to reach steady state (N) is set at 99. Total simulation time for the identification experiments described below is four times the process settling time.

In the first experiment, a binary input signal with an amplitude of 1 and switching probability of 0.5 is used. This type of input signal, which has the same spectral characteristics as white noise, was shown by Levin (1960) to be the optimal input signal for the estimation of an FIR model. The correlation matrix associated with the least squares estimates of the FIR model parameters is well-conditioned with a condition number of 10.8. Without any noise added to the process output, the corresponding estimated

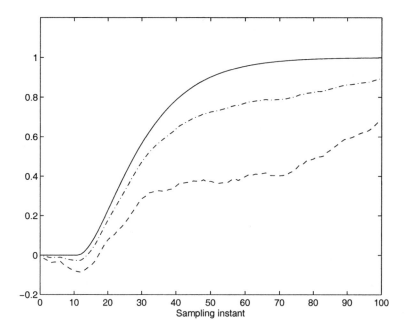

Figure 5.7: *Comparison of step responses estimated using an FIR model and a white input signal for Example 5.3 (solid: true response; dash-dotted: estimate with output noise ($\sigma = 0.1$); dashed: estimate with output noise ($\sigma = 0.3$))*

step response model obtained using Equation (5.1) is essentially identical to the true response of the process. However, with a white noise sequence of standard deviation (σ) equal to 0.1 added to the process output during the identification experiment, the estimated step response model deviates from the true process step response as seen in Figure 5.7. When the noise level is increased to a standard deviation of 0.3, the estimated step response model is seen to deviate even more from the true step response in Figure 5.7. These simulation results illustrate that, although the above input signal leads to a well-conditioned correlation matrix, the estimated step response models obtained from the FIR model parameter estimates are very sensitive to the process output noise level.

In the second experiment, we construct an input signal that is more typical of that used in an industrial setting. The particular input signal used is binary with an amplitude of 1 and consists of 6 switch lengths with a duration of about one half of the process settling time (50 samples) and 4 switch lengths with a duration of about one quarter of the process settling time (25 samples). The frequency content of this input signal is concentrated in the lower and medium frequency regions. The correlation matrix associated

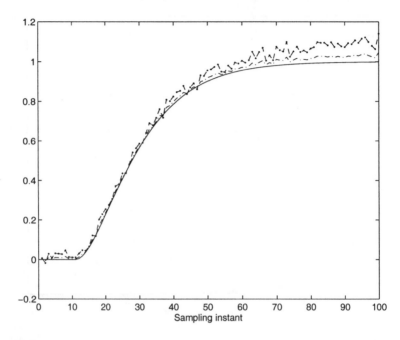

Figure 5.8: *Comparison of step responses estimated using an FIR model and an input signal with slow switches for Example 5.3 (solid: true response; dash-dotted: estimate with output noise ($\sigma = 0.1$); dashed: estimate with output noise ($\sigma = 0.3$))*

with the FIR model is ill-conditioned with a condition number of 4001. The step response estimates obtained with the same two noise levels used in the first experiment are shown in Figure 5.8. The estimated responses are much closer to the true response as compared to the results in Figure 5.7 and yet they are not smooth.

Now we study the problem further by trying to investigate the reason for the lack of smoothness in the step response estimates. We continue with the data from the second experiment. This time, the input signal is passed through 99 frequency sampling filters in parallel (corresponding to a full order FSF model structure) with centre frequencies $\frac{2\pi l}{N}$, for $l = 0, \pm 1, \ldots, \pm 49$. The filter outputs are then used to form the full order FSF correlation matrix. The diagonal elements and the row sums of the absolute values of the off-diagonal elements are shown in Figure 5.9. It is seen from these plots that the diagonal elements of the correlation matrix for the FSF structure become very small after the first 13 terms because the input signal has very little high frequency content. Also the correlation matrix is not diagonally dominant. The condition number for the correlation matrix corresponding

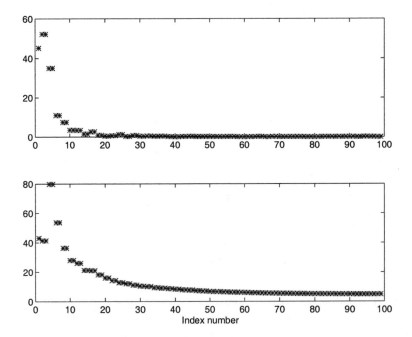

Figure 5.9: *Elements of the correlation matrix using an FSF model structure (dimension* 99×99*) for Example 5.3. Upper diagram: the diagonal elements of the correlation matrix; lower diagram: row sums of the absolute values of the off-diagonal elements*

to this full order FSF model is equal to 4001, the same as that for the FIR model. The least squares estimates of the FSF model parameters are shown in Figure 5.10. These plots show that the estimated frequency parameters are quite accurate in the low and medium frequency region despite the ill-conditioned correlation matrix, and all of the high frequency parameter estimates are poor.

This suggests that the errors in the estimated high frequency parameters are the reason for the lack of smoothness in the step response estimates. To confirm this conjecture, we estimate step response models using various reduced order FSF models. The model orders selected are $n = 99, 49, 25$ and 11. Figure 5.11 shows the estimated step response models for these four choices of n where it can be seen that, as more high frequency parameters are deleted from the estimated FSF model, the step response model becomes smoother. Also, as the number of estimated high frequency parameters is decreased, the numerical conditioning of the FSF correlation matrix improves. When $n = 49$, the condition number is 1684.7, with $n = 25$ the

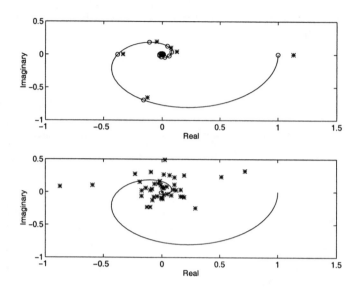

Figure 5.10: *Frequency response estimates obtained using a full order FSF model structure (N = 99) and an input signal with slow switches for Example 5.3 ('*': estimated FSF parameters; 'o': true FSF parameters; solid: true response). Upper diagram: low and medium frequency region; lower diagram: high frequency region*

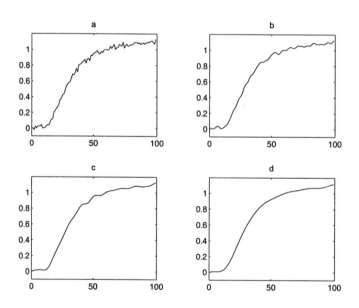

Figure 5.11: *Step response estimates obtained using various reduced order FSF models and an input signal with slow switches for Example 5.3 (a: n = 99; b: n = 49; c: n = 25; d: n = 11)*

condition number is 289.4, and with $n = 11$, the condition number is 21.0.

5.4   ERROR ANALYSIS

It has been shown in the previous section that neglecting outputs of the frequency sampling filters with centre frequencies that are not present to any significant degree in the input signal improves the numerical conditioning of the correlation matrix and results in a smoother step response estimate. However, when we delete these filters outputs, we are effectively assuming that the corresponding FSF model parameters are zero. If they are not zero, then the least squares estimates of the remaining parameters will be biased, with the amount of bias depending on the magnitudes of the neglected parameters. This section analyzes the error associated with the estimated reduced order FSF model.

Let us assume for this analysis that the data matrix Φ corresponds to the full order FSF model and has dimension $M \times N$. We will now partition Φ into Φ_n and Φ_{N-n}, where Φ_n is an $M \times n$ matrix with columns given by the n filter outputs being retained in the reduced order FSF model, and Φ_{N-n} is an $M \times (N-n)$ matrix with columns given by the $N-n$ filter outputs being neglected. Similarly, let θ, corresponding to the parameters of the full order FSF model, be partitioned into θ_n and θ_{N-n}, where θ_n contains the FSF parameters being retained and θ_{N-n} contains the neglected FSF parameters. Let us also assume that θ_n contains the lower frequency parameters and θ_{N-n} contains the higher frequency parameters. The process output Y can then be represented accordingly as

$$Y = \Phi_n \theta_n + \Phi_{N-n} \theta_{N-n} + \zeta \tag{5.14}$$

and, when using a reduced order FSF model to describe the process, the least squares estimate of the reduced order parameter vector, $\hat{\theta}_n$, is given by

$$\hat{\theta}_n = (\Phi_n^* \Phi_n)^{-1} \Phi_n^* Y \tag{5.15}$$

Substituting Equation (5.14) into Equation (5.15) leads to

$$\hat{\theta}_n = (\Phi_n^* \Phi_n)^{-1} \Phi_n^* (\Phi_n \theta_n + \Phi_{N-n} \theta_{N-n} + \zeta) \tag{5.16}$$

Thus, if the disturbance sequence ζ is zero mean, the expected value of the estimate $\hat{\theta}_n$ is

$$E[\hat{\theta}_n] = (\Phi_n^* \Phi_n)^{-1} \Phi_n^* \Phi_n \theta_n + (\Phi_n^* \Phi_n)^{-1} \Phi_n^* \Phi_{N-n} \theta_{N-n} \tag{5.17}$$

or

$$E(\hat{\theta}_n - \theta_n) = (\Phi_n^* \Phi_n)^{-1} \Phi_n^* \Phi_{N-n} \theta_{N-n} \tag{5.18}$$

Remarks:

- Equation (5.18) clearly shows that two factors determine the expected amount of error in the parameter estimates $\hat{\theta}_n$. One factor is θ_{N-n} and the other factor is the matrix $(\Phi_n^* \Phi_n)^{-1} \Phi_n^* \Phi_{N-n}$. With the FSF model structure, the magnitudes of the neglected higher frequency parameters in θ_{N-n} are generally smaller than the magnitudes of the retained lower frequency parameters in θ_n. For many chemical processes, we have found that the magnitudes of the neglected parameters are small for $n \geq 11$.

- The elements of the matrix $(\Phi_n^* \Phi_n)^{-1} \Phi_n^* \Phi_{N-n}$ are small if the elements in the matrix $\Phi_n^* \Phi_{N-n}$ are small. $\Phi_n^* \Phi_{N-n}$ contains cross-correlation terms between the outputs of the retained filters and the outputs of the neglected filters.

Now let us rewrite Equation (5.4), which converts the FSF parameters into step response coefficients, into a similar partitioned form

$$g_m = S_n(m)\theta_n + S_{N-n}(m)\theta_{N-n} \tag{5.19}$$

where $S_n(m) = [S(0,m) \ \ S(1,m) \ \ S(-1,m) \ \ \ldots \ ]$ correspond to the weighting functions defined in Equation (5.7) that are associated with the retained parameters θ_n, and $S_{N-n}(m)$ are the weighting functions corresponding to the neglected parameters. Then the error between the true step response g_m and the estimated $\hat{g}_m$ is

$$\hat{g}_m - g_m = S_n(m)(\hat{\theta}_n - \theta_n) - S_{N-n}(m)\theta_{N-n} \tag{5.20}$$

Therefore, the expected error associated with the step response estimate is

$$E[\hat{g}_m - g_m] = S_n(m)E(\hat{\theta}_n - \theta_n) - S_{N-n}(m)\theta_{N-n} \tag{5.21}$$

and substituting Equation (5.18) into Equation (5.21) gives

$$E[\hat{g}_m - g_m] = S_n(m)(\Phi_n^* \Phi_n)^{-1} \Phi_n^* \Phi_{N-n}\theta_{N-n} - S_{N-n}(m)\theta_{N-n} \tag{5.22}$$

which shows that the error in the step response estimate consists of two parts. The first term on the right-hand side of Equation (5.22) represents the error associated with the estimated FSF model parameters $\hat{\theta}_n$ and the second term comes from neglecting the higher frequency parameters θ_{N-n} in the calculation of the step response. Because the amplitudes of the weighting functions associated with the higher frequencies $S_{N-n}(m)$ are relatively small as illustrated in Figures (5.1) and (5.2), the contribution of the second term to the step response error is likely to be small for most processes with a reasonable choice for n. On the other hand, the elements of $S_n(m)$ have relatively large amplitudes, and therefore the error associated with the estimated FSF model parameters tends to dominate the error in the step response estimate.

Input Signal Design for Accurate Step Response Models

Our objective for input signal design is to reduce the expected error in the step response estimate $\hat{g}_m$ by reducing the expected error in the FSF model parameter estimates $\hat{\theta}_n$ through manipulation of the input signal's energy content. Recall the expected error given in Equation (5.18). First, the input signal chosen must have sufficient frequency content in the frequency region where the parameters are being estimated to guarantee a well-conditioned correlation matrix $\Phi_n^*\Phi_n$. Then, to reduce the cross-correlation terms which make up the matrix $\Phi_n^*\Phi_{N-n}$, the frequency content of the input signal in the frequency region where the parameters are being neglected should be small or zero.

With an estimate of the process settling time T_s, the important frequency region for parameter estimation is the low frequency region up to $\frac{\pi(n-1)}{T_s}$ radians/time, where n can be safely selected within the range of 11 to 19 for many processes. One approach would be to choose the input signal such that it contains a set of equally weighted frequencies $0, \frac{2\pi}{T_s}, \ldots, \frac{(n-1)\pi}{T_s}$. This can be realized using the multifrequency input signal design described by Schroeder (1970). This input signal consists of a sum of sinusoids with the user providing specifications for the amplitude and frequency of each sinusoid. The contribution of Schroeder (1970) was a simple formula for determining the phases of these sinusoids to minimize the overall peak factor of the resulting input signal. In principle, these types of input signals are well-suited to the identification of FSF models. However, a multisinusoidal signal is seldom used in the process industries as an input signal for process identification because it is more difficult to move the process outside of any initial nonlinearities such as valve backlash and stiction using a sinusoid-type signal, and because a binary signal is easier to implement.

One approach to a more practical input signal design is to use a clipping technique to convert the multisinusoidal signal into a binary signal. With this procedure, all points of the multisinusoidal signal greater than its average value are set to $+1$ and all points below the average value are set to -1. The following example illustrates this approach.

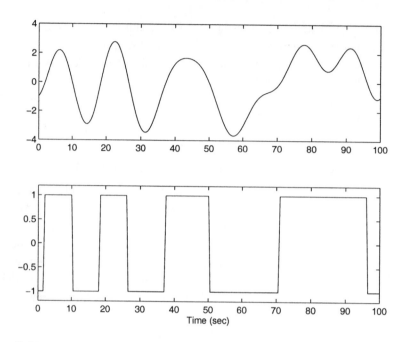

Figure 5.12: *Shroeder-phased input signal for Example 5.4. Upper diagram: multisinusoidal input signal; lower diagram: clipped input signal*

Example 5.4. Suppose that the settling time of the process to be identified is $T_s = 100$ sec with a sampling interval of $\Delta t = 0.5$ sec, and that the number of FSF model parameters to be estimated is $n = 15$. Using an amplitude of unity for each sinusoid, the corresponding Shroeder-phased multisinusoidal input signal is shown in Figure 5.12, along with the binary signal produced by applying the clipping procedure.

Another approach to input signal design for the estimation of step response models is to select the switching intervals of a binary signal according to the estimated process settling time T_s. The basic idea is that, since the first harmonic for a periodic signal with period T_s is $\frac{2\pi}{T_s}$, choosing the switching interval of a periodic binary input signal as $\frac{T_s}{2}$ will ensure that a significant amount of the input signal energy is focused at the centre frequency associated with the first pair of FSF model parameters. This idea can be extended in order to excite the higher FSF frequencies by using shorter switching intervals. For example, a general input signal may consist of a mixture of switching intervals around $\frac{T_s}{2}$, $\frac{T_s}{4}$, $\frac{T_s}{6}$ and $\frac{T_s}{8}$, which correspond to the first pair, second pair, third pair and fourth pair of FSF model parameters beyond the zero frequency. The sequence of switching intervals can be randomized in order to ensure that energy is also present in the input signal at the zero frequency.

5.5 CONFIDENCE BOUNDS FOR FREQUENCY RESPONSE AND STEP RESPONSE ESTIMATES

The statistical properties for the least squares estimate $\hat{\theta}$ of the FSF model parameters given in Equation (4.26) are used in this section to develop statistical confidence bounds for frequency response and step response models estimated using the FSF model structure. We begin by stating the key assumptions on which all of this analysis is based and then summarize the key properties of the least squares estimator, which may be found in Ljung (1987).

We assume for the process being described in Equation (4.25) that:

A 5.1 The process is stable, linear and time invariant with finite settling time T_s and the parameter N is chosen to be greater than or equal to $\frac{T_s}{\Delta t}$.

A 5.2 The disturbance ζ is zero mean, normally distributed white noise with variance σ^2.

A 5.3 $n = N$, or n is chosen such that the neglected frequency parameters are negligible in magnitude relative to the parameters being retained in the model.

Based on these assumptions, we can state the following properties of the least squares estimate $\hat{\theta}$ obtained from Equation (4.26):

- **Bias:** The estimate $\hat{\theta}$ is unbiased, i.e. $E[\hat{\theta}] = \theta$.

- **Variance:** The covariance of the parameter estimates is given by

$$E[(\hat{\theta} - \theta)(\hat{\theta} - \theta)^*] = (\Phi^*\Phi)^{-1}\sigma^2 \tag{5.23}$$

- **Distribution properties:** Since only a linear operation is involved in estimating the parameters, $\hat{\theta}$ will follow a normal distribution.

If the variance, σ^2, of the disturbance is unknown, then a consistent estimate of σ^2 can be obtained from

$$\hat{\sigma}^2 = \frac{1}{M - n - 1}(Y - \Phi\hat{\theta})^T(Y - \Phi\hat{\theta}) \tag{5.24}$$

Note that the normal distribution of the parameter estimates $\hat{\theta}$ is contingent on the data length M being sufficiently large. For short data lengths, the estimates follow instead a student-T distribution (Goodwin and Payne, 1977) and converge to the normal distribution as the data length increases. It can be seen from Kreyszig (1988) that this will occur when the degrees of freedom associated with the estimate of the noise variance $(M - n)$ exceeds 100.

Confidence Bounds in the Frequency Domain

Theorem 5.1: Suppose that assumptions A 5.1-A 5.3 are satisfied. Then the distance between the true and the estimated frequency response at $w_l = \frac{2\pi l}{N}$, $l = 0, \pm 1, \pm 2, \ldots, \pm \frac{n-1}{2}$, is bounded by

$$|\hat{G}(e^{j\frac{2\pi l}{N}}) - G(e^{j\frac{2\pi l}{N}})| \leq p\sigma_{\theta,l} \qquad (5.25)$$

with probability $P(p)$, where $\sigma_{\theta,l}$ is the standard deviation associated with $\hat{G}(e^{j\frac{2\pi l}{N}})$, found by taking the square root of the corresponding diagonal element in Equation (5.23), and $P(1) = 0.683$, $P(2) = 0.954$ and $P(3) = 0.997$ according to the specified level of the normal distribution.

Proof: Assumptions A 5.1-A 5.3 guarantee that the least squares estimate $\hat{\theta}$ is unbiased and follows a normal distribution. Given that the parameters being estimated correspond to the process frequency response at $w_l = \frac{2\pi l}{N}$ for $l = 0, \pm 1, \pm 2, \ldots, \pm \frac{n-1}{2}$, then the result follows directly from the variance of the individual parameter estimates.

Equation (5.25) tells us that the distance between the true and the estimated frequency response at $w_l = \frac{2\pi l}{N}$ will be less than $p\sigma_{\theta,l}$ with probability $P(p)$. Graphically, this means that, on the complex plane, if $\hat{G}(e^{j\frac{2\pi l}{N}})$ is used as the centre of a circle and $p\sigma_{\theta,l}$ is used as the radius of this circle, then the true process frequency response $G(e^{j\frac{2\pi l}{N}})$ must lie within this circle with a probability of $P(p)$. As for the rest of the frequencies, an interpolation technique is used. Let the estimated process frequency response $\hat{G}(e^{jw})$ be represented by

$$\hat{G}(e^{jw}) = H(e^{jw})\hat{\theta} \qquad (5.26)$$

where

$$H(e^{jw}) = [H^0(e^{jw}) \ \ H^1(e^{jw}) \ \ H^{-1}(e^{jw}) \ldots H^{\frac{n-1}{2}}(e^{jw}) \ \ H^{-\frac{n-1}{2}}(e^{jw})]$$

with

$$H^l(e^{jw}) = \frac{1}{N} \frac{1 - e^{-jwN}}{1 - e^{j\frac{2\pi l}{N}}e^{-jw}} \qquad (5.27)$$

being the frequency response of the *lth* filter. The following theorem presents the bounds within which the true frequency response lies at any arbitrary frequency.

Theorem 5.2: Suppose that assumptions A 5.1-A 5.3 are satisfied. Then the distance between the true and the estimated frequency response at w, $-\pi \leq w \leq \pi$, is bounded by

$$|\hat{G}(e^{jw}) - G(e^{jw})| \leq p \times \Sigma(e^{jw}) \qquad (5.28)$$

with probability $P(p)$, where

$$\Sigma(e^{jw})^2 = H(e^{jw})(\Phi^*\Phi)^{-1}H^*(e^{jw})\sigma^2 \qquad (5.29)$$

and $P(1) = 0.683$, $P(2) = 0.954$ and $P(3) = 0.997$ according to the specified level of the normal distribution.

Proof: Under assumption A 5.1, it is valid to represent the true process frequency response $G(e^{jw})$ as

$$G(e^{jw}) = H(e^{jw})\theta \qquad (5.30)$$

Assumptions A 5.1-A 5.3 guarantee that $\hat{G}(e^{jw})$ is an unbiased estimate of $G(e^{jw})$, i.e. $E[\hat{G}(e^{jw})] = G(e^{jw})$ and that $\hat{G}(e^{jw})$ follows a normal distribution since it is obtained from a linear transformation of the normally distributed parameter estimate $\hat{\theta}$. We can compute the variance of the estimated process frequency response at a specific frequency w from the covariance of the parameter estimates as follows. First we can write

$$\hat{G}(e^{jw}) - G(e^{jw}) = H(e^{jw})(\hat{\theta} - \theta) \qquad (5.31)$$

followed by

$$(\hat{G}(e^{jw})-G(e^{jw}))(\hat{G}(e^{jw})-G(e^{jw}))^* = H(e^{jw})(\hat{\theta}-\theta)(\hat{\theta}-\theta)^*H^*(e^{jw}) \quad (5.32)$$

Hence,

$$E[|\hat{G}(e^{jw}) - G(e^{jw})|^2] = H(e^{jw})E[(\hat{\theta} - \theta)(\hat{\theta} - \theta)^*]H^*(e^{jw})$$

which, by substituting Equation (5.23), leads to

$$E[|\hat{G}(e^{jw}) - G(e^{jw})|^2] = H(e^{jw})(\Phi^*\Phi)^{-1}H^*(e^{jw})\sigma^2 = \Sigma(e^{jw})^2 \qquad (5.33)$$

Therefore, from the properties of a normal distribution, we have

$$|\hat{G}(e^{jw}) - G(e^{jw})| \le p \times \Sigma(e^{jw}) \tag{5.34}$$

with probability $P(p)$.

Equation (5.28) indicates that the true process frequency response $G(e^{jw})$ lies inside the circle on the complex plane, centred at $\hat{G}(e^{jw})$ with radius equal to $p \times \Sigma(e^{jw})$.

Confidence Bounds for the Step Response Model

We have also been able to obtain confidence bounds for the step response model derived from the FSF model using Equation (5.4). The basic idea is to represent the step response coefficients as a linear transformation of the estimated FSF parameters, and then map the covariance matrix from the frequency domain to the time domain.

Theorem 5.3: Let the estimated step response be represented by

$$\hat{g}_m = S(m)\hat{\theta} \tag{5.35}$$

where $S(m) = [S(0,m) \quad S(1,m) \quad \ldots \quad S(-\frac{n-1}{2}, m)]$ with $S(l,m)$ defined in Equation (5.7). Then, under assumptions A 5.1-A 5.3, the error between the true process step response weight g_m and the estimated step response weight $\hat{g}_m$ is bounded by

$$|\hat{g}_m - g_m| \le p \times \delta(m) \tag{5.36}$$

with probability $P(p)$, where $\delta(m)$ is given by

$$\delta(m)^2 = S(m)(\Phi^*\Phi)^{-1}S^*(m)\sigma^2 \tag{5.37}$$

Proof: Under assumption A 5.1, the true process step response can be represented by

$$g_m = S(m)\theta \tag{5.38}$$

From assumptions A 5.1-A 5.3, we know that $\hat{\theta}$ is an unbiased and normally distributed estimate of θ. Therefore, $\hat{g}_m$ is an unbiased and normally distributed estimate of g_m. The variance of the estimated step response coefficient at the sampling instant m is given by

$$
\begin{aligned}
E[(\hat{g}_m - g_m)^2] &= E[S(m)(\hat{\theta} - \theta)(\hat{\theta} - \theta)^*S^*(m)] \\
&= S(m)(\Phi^*\Phi)^{-1}S^*(m)\sigma^2 \\
&= \delta(m)^2
\end{aligned} \tag{5.39}
$$

and the bounds in Equation (5.36) follow directly.

Applying Equation (5.36), the trajectory of the true step response g_m for $m = 0, 1, \ldots, N - 1$ lies inside the envelope given by $\hat{g}_m \pm p \times \delta(m)$ with probability $P(p)$. This envelope provides the confidence bound on the estimated step response model.

5.6 GENERALIZED LEAST SQUARES ALGORITHM

In this section, we present an iterative algorithm in the spirit of the generalized least squares approach (Goodwin and Payne, 1977), for simultaneous estimation of an FSF process model and an autoregressive (AR) noise model. The unique features of our algorithm are the application of the $PRESS$ statistic introduced in Chapter 3 for both process and noise model structure selection to ensure whiteness of the residuals, and the use of covariance matrix information to derive statistical confidence bounds for the final process step response estimates. An important assumption in this algorithm is that the noise term $\xi(k)$ can be described by an AR time series model given by

$$\xi(k) = \frac{1}{F(z)}\epsilon(k) \tag{5.40}$$

where $F(z) = 1 + f_1 z^{-1} + \cdots + f_m z^{-m}$ and $\{\epsilon(k)\}$ is assumed to be a zero mean, white noise sequence.

The algorithm will be presented as a step-by-step procedure for identification of a MISO system. The user must first provide estimates for the times to steady state for the individual subsystems given by N_i, $i = 1, 2, \ldots, p$, the maximum values to be considered for the reduced model orders n_i, $i = 1, 2, \ldots, p$, and the maximum noise model order m.

Step 1: Assume $F(z) = 1$.

Step 2: Prefilter the process inputs and output with $F(z)$, namely $u_{i,f}(k) = F(z)u_i(k)$ for $i = 1, \ldots, p$ and $y_f(k) = F(z)y(k)$.

Step 3: Determine least squares estimates for the FSF process model parameters using the $PRESS$ statistic to select the model order n_i for each subsystem.

Step 4: Estimate the noise sequence $\hat{\zeta} = Y - \Phi\hat{\theta}$.

Step 5: From $\hat{\zeta}$, estimate the noise model $F(z)$ using the least squares method along with the $PRESS$ statistic to determine the value for m.

Step 6: Check for convergence of the FSF model parameter estimates $\hat{\theta}$. If the parameters have converged, go to Step 7. Otherwise, return to Step 2 with the updated estimate of $F(z)$.

Step 7: Calculate the covariance of the parameter estimates and use it to calculate the statistical confidence bounds for the estimated step response models.

5.7 INDUSTRIAL APPLICATION: IDENTIFICATION OF A REFINERY DISTILLATION TRAIN

5.7.1 Process description

The process used in this study is part of a distillation train in the BTX (Benzene, Toluene, Xylene) plant at Sunoco's refinery in Sarnia, Canada. The process flow diagram of the complete distillation train is illustrated in Figure 5.13. The plant consists of four distillation columns: Benzene, Toluene, mixed-Xylene (MX) and ortho-Xylene (OX). The bottoms product of the first three columns feed their respective downstream columns. Material is balanced in each of the columns by controlling the bottoms level through manipulation of the outlet flow. This means that the manipulated variable for controlling the bottoms level in one column is a disturbance to the following tower. The scope of the identification problem studied here includes the final two columns (MX and OX towers) of the distillation train, with the objective being to develop step response models for the design of a multivariable model predictive controller. Figure 5.14 is a more detailed schematic of these two columns. The operation of the process with reference to Figure 5.14 is described below.

Feed from the Toluene tower is preheated (1) by the MX tower distillate product and then enters the MX tower (2) at approximately the middle tray. The MX tower overhead is totally condensed using heated water from the gas fired reboiler (3). Steam is produced in the condenser (4) and delivered to a utility header. Flow of water into the shell side of the condenser is manipulated by a level controller. The condensed MX tower overhead material is collected in an accumulator (5). A level controller manipulates the reflux to the tower to maintain the accumulator level. The overhead product flow is set externally, either by an advanced controller or manually. At the bottom of the MX tower the flow is divided into two streams. One of the streams is circulated through the gas fired reboiler, where it is partially vaporized and then returned to the MX tower. The vaporization rate is set

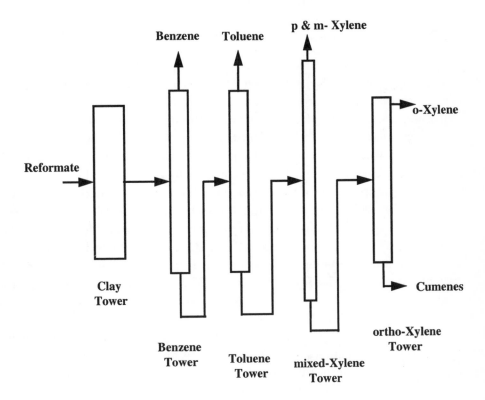

Figure 5.13: *Process flow diagram for distillation train*

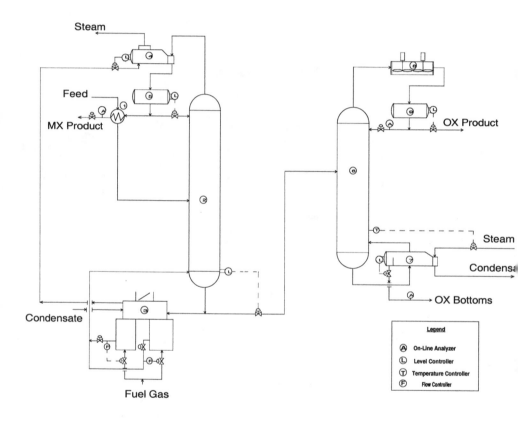

Figure 5.14: *Process schematic of MX and OX distillation towers*

externally and is cascaded to the fuel gas flow. The second stream feeds the OX tower (6) with the flow manipulated by the MX bottoms level controller. The reboil heat for the OX tower is provided by high pressure steam. Steam flow to the reboiler (7) is controlled by a temperature controller used to maintain a specified bottom tray temperature. Flow of the OX tower bottoms product is controlled by a level controller for the reboiler. An air fin cooler (8) is used to totally condense the OX tower overhead material and is subsequently collected in an accumulator (9). A level controller manipulates the OX tower product flow to maintain the level in the accumulator. Unlike the MX tower, the reflux flow for this tower is set externally.

The following variables were selected as the dependent variables (process outputs):

y_1: o-Xylene product purity.

y_2: Cumenes concentration in o-Xylene product.

y_3: o-Xylene concentration in OX tower bottoms.

The independent variables (process inputs), with the first three being designated as manipulated variables, were then selected as:

u_1: Overhead MX tower product flow; this variable indirectly manipulates the reflux flow to the MX tower through the accumulator level controller.

u_2: OX tower reflux flow.

u_3: Temperature setpoint at the bottom of OX tower; determines the steam flow to the reboiler.

u_4: Feed flow to the MX tower; a feedforward variable manipulated by a level controller on the Toluene tower.

u_5: Vaporization rate in the gas fired reboiler; a feedforward variable manipulated by a separate controller for the reboiler.

5.7.2   Dynamic response testing

Following selection of the independent and dependent variables, a pre-test of the unit's regulatory control system was conducted. The objective of the pre-test was to collect information on the settling time of the process and the tuning of the regulatory controllers. Since the multivariable model predictive controller was to be eventually superimposed over the basic regulatory

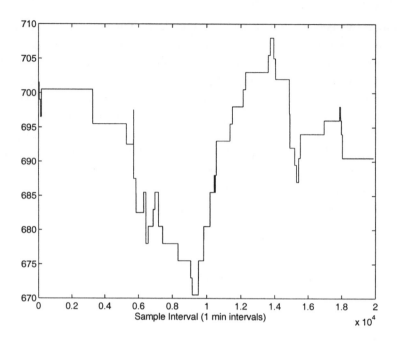

Figure 5.15: *Typical input perturbation signal*

controllers, it was essential that these loops be well tuned. The pre-test also gave information on the quality of process signals and the size of step moves required to generate a sufficient response in the dependent variables.

Dynamic testing of the plant was conducted on a continuous basis over a 14 day period. During this test each independent variable was perturbed and the process data was recorded at one minute intervals. The selection of which variable to move and the direction of movement was made on the basis of maintaining products at specification and avoiding saturation of the underlying regulatory PID controllers.

In addition to maintaining product specifications, it was essential to use different corrective actions in response to a particular control situation in order to avoid correlation between the independent variables. The control engineer on shift was responsible for ensuring randomness of the test. The engineer also recorded unusual occurrences that might impact negatively the test data. Examples of such events include heater upsets, analyzer and equipment failures and data acquisition errors. Approximately 20 moves were made in each independent variable. An example of a perturbation signal is presented in Figure 5.15.

5.7.3   Results

The response test data was divided into 16 minutes intervals, and subsequently averaged over the intervals. This resampling reduced the amount of data and was justified because the concentration analyzer measurements were updated only every 15 minutes.

As mentioned above, a number of unusual events may occur during the response test. One way to exclude this data from the analysis is by differencing the input and output data. This permits identification data to be selectively removed, without invalidating the entire response test. Another advantage of differencing is that the process does not have to be at steady state for the response test to start. This is very useful for industrial processes that often operate in a continuous dynamic state because of disturbances that drive the process away from product specifications.

The remaining subsections will examine the results obtained using the FSF approach, FIR models obtained using the least squares method, and models obtained using DMI, a commercially available process identification software package. With this process, there are a total of 15 input-output relationships to be estimated. For brevity, we will only examine a subset of the results to highlight the features of the FSF approach. One key difference between the various approaches that needs to be mentioned at the outset is that the initial value of each step response model (g_0) has been estimated with the FSF approach but has been set equal to zero with the FIR and DMI approaches.

DMI is a commercial product of DMC Corporation (Cutler and Yocum, 1991). The selection of DMI models is an interactive process that involves analyzing results for a number of different times to steady state. Smooth and non-smooth step response models, which in the latter case correspond to step response models generated from least squares estimated FIR models, are presented graphically to the user for each input-output pair. A proprietary smoothing method is used to reduce the effect of noise on the non-smooth models while minimizing the residual errors in the fit of the data. A particular model is selected for a controller application if there is a reasonable match between the smooth and non-smooth curves, and if the response appears to have reached a steady state. The selected times to steady state for this system are summarized in Table 5.1.

Estimates of the times to steady state are required for application of the FSF identification method. In order to compare with the DMI results, we have chosen the same times to steady state presented in Table 5.1. The reduced FSF process model orders that were selected through application

	u_1	u_2	u_3	u_4	u_5
y_1	1080	240	240	960	360
y_2	540	240	240	1080	480
y_3	960	240	240	960	480

Table 5.1: *Estimated times to steady state (in minutes)*

	u_1	u_2	u_3	u_4	u_5
y_1	15	1	3	11	5
y_2	7	13	3	9	7
y_3	5	7	7	11	5

Table 5.2: *Selected FSF model orders*

of the iterative algorithm presented in Section 5.6 are summarized in Table 5.2.

5.7.4 Use of *PRESS* for model structure selection

One of the key features of the proposed generalized least squares algorithm is the use of the *PRESS* for both process and noise model order selection. Figure 5.16 provides a graphical representation of the number of terms selected by the *PRESS* statistic for each of the five subsystems associated with the first output variable y_1. The amount of reduction in the *PRESS* by each input variable and the number of terms selected are felt to be measures of the quality of the input signal design, the signal-to-noise ratio and the existence of an input-output relationship. For example, Figure 5.16 shows that u_2 and u_3 do not contribute significantly to the reduction of the *PRESS* for y_1. More specifically, the *PRESS* is indicating that there is no relation between u_2 and y_1 because even the addition of a single term causes the *PRESS* to increase. Although three terms have been selected for the u_3 to y_1 relationship, the *PRESS* was only reduced slightly.

The *PRESS* clearly indicates that the correlations between these two inputs (u_2 and u_3) and the process output (y_1) in this particular data set are weak. However, from a physical point of view, it is expected that these two input variables, OX tower reflux flow and temperature setpoint at the bottom of the OX tower, should have an effect on y_1, the OX tower product purity. One possible explanation is that the settling times were overestimated for these two relationships (240 minutes each) combined with the

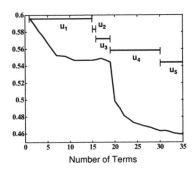

Figure 5.16: *PRESS for process output* y_1

fact that most switches in the input signals were spaced much larger than the estimated settling times. This would mean that not enough information was present in the data to identify the dynamic relationships. In general, if the model order selected by the *PRESS* is low (e.g. $n < 7$) for a given input-output pair, this may be an indication that the input signal lacks excitation in the medium frequency region (e.g. switches ranging from $\frac{N_i \Delta t}{8}$ to $\frac{N_i \Delta t}{2}$). If the switching frequency is increased but the model order remains low, this is possibly an indication that there is either no relationship between this input-output pair or that the overall signal-to-noise ratio is very low. It is worth noting that the DMI analysis also indicated that there were no significant relationships between these variables in this data set.

5.7.5 Use of noise models to remove feedback effects

During plant tests, the input variables are sometimes adjusted by the operator in order to maintain the product at its specification. Different corrective actions are taken in response to a particular control situation in order to avoid correlation between the independent variables. However, this type of operator intervention introduces feedback into the test data that can lead to significant bias in the estimated process models. To remove this effect, noise

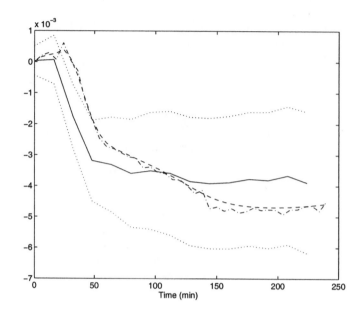

Figure 5.17: *Step response models relating u_2 to y_2 (solid: FSF; dotted: 99% confidence bounds on FSF model; dashed: DMI (smooth); dash-dotted: FIR)*

models must be built and used in the design of prefilters for the input-output data used to estimate process models (MacGregor and Fogal, 1995). The FIR and DMI models assume a noise model where $F(z) = (1 - z^{-1})$. Our results show that the noise models are more complicated than this simple model. The biasing effect on the FIR and DMI models can appear as an inverse response because the initial value of the step response has been set equal to zero in these cases. For example, let us examine the results for the u_2 to y_2 relationship presented in Figure 5.17. The FIR and DMI estimated step responses clearly show an inverse response. However, the FSF results indicate a time delay of approximately 20 minutes, which is believed to be closer to the actual process behaviour.

5.7.6   Use of confidence bounds for judging model quality

Another important role of the noise model in the iterative algorithm is to ensure whiteness of the residuals. This allows us to estimate the covariance of the FSF model parameter estimates and then to develop statistical confidence bounds for the corresponding step response estimates. In order to apply these results, it is important that the bias error in the model arising due to unmodelled dynamics be small relative to the variance error caused by the presence of noise in the measured outputs. With our approach, we believe that the model order selected by the *PRESS* can be used as an indication of whether this condition holds. For instance, a low model order (e.g. $n_i = 3$) indicates that the bias error is significant leading to a biased step response estimate. In this case, the bounds would not be expected to enclose the true step response and could only be used as a crude measure of possible responses. On the other hand, a higher order model (e.g. $n_i > 5$) indicates that the bias error is relatively small compared to the variance error and, in this case, the bounds would be expected to enclose the true response at the confidence level chosen.

For example, Figure 5.18 shows the estimated step responses for u_3 to y_2. The FSF model order selected for this input-output pair was 3 which means that the true response may not lie within the confidence bounds. Evidence for this is found by observing that the confidence bounds at time zero do not include zero as a possible value. However, for most processes encountered in the process industries, the initial value of the step response is known to be equal to zero. By comparison, for higher order models, the confidence bounds seem to always enclose zero as a possible value at time zero. For example, Figure 5.19 shows the estimated step responses for u_1 to y_1 that has a selected model order of 15. In this case, the bounds are believed to enclose the true response with a 99% confidence level.

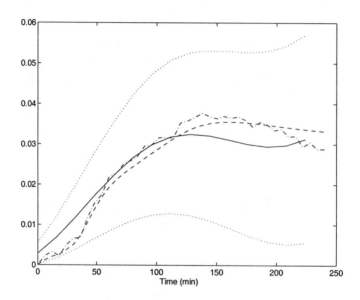

Figure 5.18: *Step response models relating u_3 to y_2 (solid: FSF; dotted: 99% confidence bounds on FSF model; dashed: DMI (smooth); dash-dotted: FIR)*

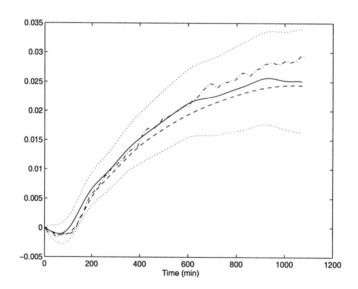

Figure 5.19: *Step response models relating u_1 to y_1 (solid: FSF; dotted: 99% confidence bounds on FSF model; dashed: DMI (smooth); dash-dotted: FIR)*

Chapter 6

New Frequency Domain PID Controller Design Method

6.1 INTRODUCTION

This chapter presents a new PID controller design method based on process frequency response information. The novel ideas lie in the way that the closed-loop performance is specified via the desired response of the control signal and in the use of only two process frequency response points in the design. The relationship between the process frequency response and its step response, developed in Chapter 5, is exploited here to determine the frequency information to be used for controller design.

This chapter consists of seven sections. In Section 6.2, we present the desired closed-loop performance specification in terms of the controller output response for both stable and integrating processes. In order to gain some insight into the new design method, Section 6.3 discusses a least squares approach to the PID controller parameter solutions. From the least squares solution, we propose in Section 6.4 a simpler approach based on process information at just two frequency points. Section 6.5 addresses the question of which two frequencies should be used in the design. Section 6.6 discusses the choice of a PI or a PID controller and provides some guidance with respect to the selection of the closed-loop performance parameters. Extensive simulation studies are performed in Section 6.7 and the results are compared with other design methods.

Portions of this chapter have been reprinted from *IEE Proceedings-Control Theory and Applications* **142**, L. Wang, T.J.D. Barnes and

W.R. Cluett, "New frequency-domain design method for PID controllers", pp. 265-271, 1995, with permission from IEE.

6.2   CONTROL SIGNAL SPECIFICATION

One of the most common features of many PID controller designs is that performance is specified in terms of the trajectory of the desired closed-loop process output response to a setpoint change. Here, we propose to specify the closed-loop performance in terms of the desired behaviour of the controller output or control signal in response to a setpoint change.

Consider the feedback system illustrated in Figure 6.1, where u and y are the control signal and measured process output, respectively, r is the setpoint, d is the load disturbance, and C and G denote the controller and plant transfer functions, respectively. We will assume that C has the structure of a PID controller given by

$$C(s) = \frac{c_2 s^2 + c_1 s + c_0}{s} \tag{6.1}$$

$$= K_c(1 + \frac{1}{\tau_I s} + \tau_D s) \tag{6.2}$$

or that of a PI controller given by

$$C(s) = \frac{c_1 s + c_0}{s} \tag{6.3}$$

where the **proportional gain** is

$$K_c = c_1 \tag{6.4}$$

the **integral time constant** is

$$\tau_I = \frac{c_1}{c_0} \tag{6.5}$$

and the **derivative time constant** is

$$\tau_D = \frac{c_2}{c_1} \tag{6.6}$$

The transfer function from the setpoint r to the control signal u is given by

$$G_{r \to u}(s) = \frac{C(s)}{1 + C(s)G(s)} \tag{6.7}$$

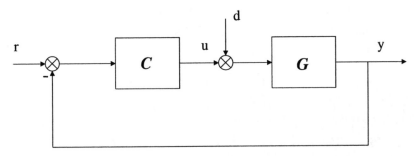

Figure 6.1: *Block diagram of the feedback control system*

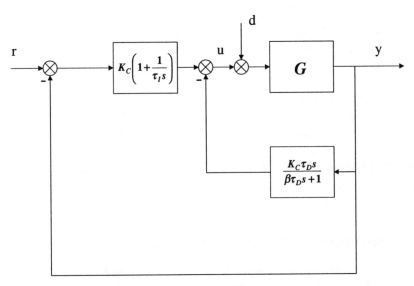

Figure 6.2: *Block diagram of a practical PID controller implementation*

The transfer function from the setpoint r to the process output y is given by

$$G_{r \to y}(s) = \frac{C(s)G(s)}{1 + C(s)G(s)} \tag{6.8}$$

and the transfer function from the load disturbance d to the control signal u is given by

$$G_{d \to u}(s) = -\frac{C(s)G(s)}{1 + C(s)G(s)} \tag{6.9}$$

Our approach is to specify the trajectory of the control signal in response to a setpoint change in order to take advantage of the characteristics of Equation (6.7). The reasons we choose to work with $G_{r \to u}$ instead of either $G_{r \to y}$ or $G_{d \to u}$ are:

- the behaviour of the control signal is an important consideration when assessing overall performance of a PID controller in a process control application because it is often desirable or necessary for the control signal to have a smooth response;

- the transfer function that governs the control signal response to a setpoint change, $G_{r \to u}$, is in some sense less dependent on the process dynamics, in that $G(s)$ does not appear in the numerator as it does in both $G_{r \to y}$ and $G_{d \to u}$. Therefore, a specification on $G_{r \to u}$ can be given without requiring detailed knowledge about the process dynamics, such as the process delay and process zeros.

Our expectation is that if the control signal responds in a smooth manner, the resulting process output response $G_{r \to y}$ will also be smooth. Because $G_{d \to u}$ is simply the negative of $G_{r \to y}$, we believe our specification on $G_{r \to u}$ will ensure both a smooth response in the process output to a setpoint change and a smooth control signal response to a disturbance.

We now present our PID controller design specifications for two types of processes frequently encountered in the process industries.

6.2.1   Specification for stable processes

The type of control signal response to a step setpoint change encountered with a stable process under feedback control would be familiar to a process engineer. The first key feature is the immediate step change in the control signal when the setpoint value is changed. It is common practice to place the derivative term of the PID controller in the feedback loop so that it only acts on the filtered process variable to avoid derivative kick (see Figure 6.2).

Thus, the initial step change in the manipulated variable is solely due to the proportional control action. The second important feature is that because of the integral action and the assumed stability of the closed-loop system, the control signal trajectory, in terms of the deviation from its initial steady state value, exponentially converges to $\frac{\bar{r}}{K}$, where K is the steady state gain of the process and $\bar{r}$ is the value of the setpoint change. For processes with time delay, the controller output does not begin to converge to its final value until a time period equal to the process delay has passed.

Bearing in mind these two key features, our objective is to develop a mathematical description of the desired control signal response to a step setpoint change. Let us begin by introducing two new design parameters: α, which is related to the desired initial change in the control signal for a given step setpoint change, and τ, the desired time constant for the exponential response of the control signal following the initial change. The desired trajectory of $u(t)$ for a step setpoint change of $\bar{r}$ can then be described mathematically as

$$u(t) = \frac{\bar{r}}{K}[\alpha + (1 - \alpha)(1 - e^{-\frac{t}{\tau}})] \tag{6.10}$$

Figure 6.3 illustrates the desired control signal response to a unit step setpoint change for $\alpha = 0.5, 1, 2$, with a normalized time constant $\tau = 1$ and a steady state process gain $K = 1$. It can be readily shown that $u(0) = \frac{\bar{r} \times \alpha}{K}$ and $u(\infty) = \frac{\bar{r}}{K}$. Therefore, the parameter α determines the initial change in the control signal expressed as a fraction of the total change required to achieve the new setpoint. It also determines the relative response speed between the open-loop process and the desired closed-loop system. For instance, when $\alpha = 1$, the speed of the desired closed-loop system is equal to the open-loop process response. When $\alpha < 1$, the speed of the desired closed-loop system is slower than the open-loop response and when $\alpha > 1$, the speed of the desired closed-loop system is faster than the open-loop response. Typical values of α might be in the range 0.25 to 1.5, which correspond to initial changes in the control signal of 25% to 150% relative to the final steady state change.

The parameter τ determines the speed of convergence of the control signal to its steady state value. We show later that the choice of the parameter τ can be related to α, meaning that we can reduce the number of design parameters from two to one.

With the specification on the trajectory of the desired control signal given by Equation (6.10), the transfer function relating the setpoint to the

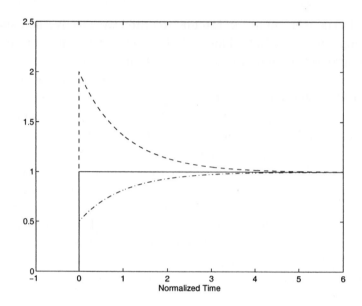

Figure 6.3: *Control signal trajectories in response to a unit step setpoint change (solid: $\alpha = 1$; dash-dotted: $\alpha = 0.5$; dashed: $\alpha = 2$)*

desired control signal is given by

$$
\begin{aligned}
G_{r \to u}(s) &= \frac{C(s)}{1 + C(s)G(s)} \\
&= \frac{1}{K} \frac{\alpha \tau s + 1}{\tau s + 1}
\end{aligned}
\tag{6.11}
$$

and the desired closed-loop transfer function relating the setpoint to the process output is given by

$$
G_{r \to y}(s) = G_{r \to u}(s)G(s) = \frac{1}{K} \frac{\alpha \tau s + 1}{\tau s + 1} G(s)
\tag{6.12}
$$

We can see that, with the control signal specification in Equation (6.11), a lead-lag element has been added in series to the open-loop process transfer function $G(s)$ to form the desired closed-loop transfer function in Equation (6.12).

As an aside, this approach to controller design can be considered from a pole-placement point of view. For a stable process under PID control, the only pole in the open-loop system that needs to be moved to achieve closed-loop stability is the one located at the origin of the complex plane introduced by the integrator in the controller. Therefore, we can think of the proposed design method as an attempt to directly specify the closed-loop pole locations, with the open-loop pole located at the origin moved to $-\frac{1}{\tau}$ and the remaining open-loop poles left untouched.

Choice of the Design Parameters

With respect to the problem of choosing α and τ for the control signal performance specification, we recommend that their values be related to the process dynamics in order to achieve a desired closed-loop response. Three examples are presented here to show how to choose α and τ in such a way as to cancel the dominant process pole using the zero of the lead-lag element in Equation (6.11).

A. First order plus delay model

$$G(s) = \frac{e^{-ds}}{Ts + 1} \tag{6.13}$$

Here we will let $\alpha\tau = T$ in order to cancel the single pole in the transfer function $G(s)$, which gives $\tau = \frac{T}{\alpha}$. Then we can choose α according to the desired closed-loop response speed. For instance, a larger value of α will result in a larger initial change in the control signal and, in turn, both a faster control signal response and process output response. From a pole-placement point of view, a larger value of α means a shift to the left of the closed-loop pole $-\frac{1}{\tau}$.

B. Laguerre model

$$G(s) = \sum_{i=1}^{N} c_i L_i(s) \tag{6.14}$$

In Chapter 2, it was stated that if the process is greater than first order but without time delay, a reasonable choice for the scaling factor p can be based on the dominant time constant of the process. In this case, we can let $\alpha\tau = \frac{1}{p}$ to cancel this dominant pole in $G(s)$, which gives $\tau = \frac{1}{\alpha p}$ allowing us to choose α to bring about the desired closed-loop response speed.

C. Step response model

Suppose we have a step response of the process from which we can obtain an estimate of the process settling time T_s and time delay d. In this case, we can crudely approximate the response by that of a first order plus delay process and select $\alpha\tau = \frac{T_s - d}{5}$ or $\tau = \frac{T_s - d}{5\alpha}$. Then, we can vary α to determine the closed-loop response speed.

In some cases, a choice of $\alpha = 1$, which corresponds to the specification of a step change in the control signal and therefore a closed-loop response speed equal to the open-loop response speed, may be desirable. From Equation (6.12), we can see that this choice of α does not require a choice for the parameter τ.

6.2.2   Specification for integrating processes

The control signal trajectories for integrating processes under PID control are different from those used for stable processes. As with the stable case, the control signal will have an initial change due to the proportional action in the controller but then, with integrating processes, the signal will exponentially converge to zero due to the presence of the double integrator in the open-loop transfer function, assuming stability of the closed-loop system. In this case, the transfer function from the setpoint to the control signal is required to be at least second order.

There are also some other characteristics that the closed-loop system must have when the open-loop transfer function contains a double integrator. Suppose that the process transfer function $G(s)$ has the following form, with the stable part represented by $H(s)$

$$G(s) \;=\; \frac{1}{s}H(s) \tag{6.15}$$

$$=\; \frac{K}{s}[1 + \gamma_1 s + \cdots] \tag{6.16}$$

where K is the steady state value of the stable part of the transfer function and γ_1 is the second coefficient in the Taylor series expansion of the stable part after factoring out K. It can be shown that, for an integrating process under PID control, the moment expansion of the closed-loop transfer function $G_{r \to y}$ has the form

$$G_{r \to y}(s) = 1 + 0s + \beta_2 s^2 + \cdots \tag{6.17}$$

where the first coefficient is unity and the second coefficient is zero. These characteristics must be reflected in the specification of the control signal trajectories in order to arrive at a successful design.

Suppose we choose the desired closed-loop transfer function relating the setpoint to the control signal to be of the form

$$G_{r \to u}(s) = \frac{s}{K}\frac{(2\zeta\tau - \gamma_1)s + 1}{\tau^2 s^2 + 2\zeta\tau s + 1} \tag{6.18}$$

where the time constant τ and the damping factor ζ are the design parameters, and K and γ_1 are the process parameters defined in Equation (6.16). Combining Equation (6.18) with the process transfer function $G(s)$ gives a desired closed-loop transfer function in the following form

$$G_{r \to y}(s) = G_{r \to u}(s)G(s) = \frac{1}{K}\frac{(2\zeta\tau - \gamma_1)s + 1}{\tau^2 s^2 + 2\zeta\tau s + 1}H(s) \qquad (6.19)$$

where it can be shown through polynomial division that $G_{r \to y}(s)$ satisfies Equation (6.17).

Types of Integrating Processes

We now divide the general class of integrating processes described by Equation (6.16) into two types based on the sign of the parameter γ_1. In both cases, the trajectory of the desired control signal is scaled with respect to $|\gamma_1|$. A single tuning parameter β ($\beta > 0$) is introduced such that the time constant of the desired control signal response is given by $\tau = \beta|\gamma_1|$. This makes analysis and tuning straightforward.

Type A: $\gamma_1 < 0$

We refer to this type of integrating process as being lag dominant, which represents the majority of integrating processes encountered in the process industries. The time constant of the desired control signal is chosen as

$$\tau = \beta|\gamma_1| \qquad (6.20)$$

Then Equation (6.18) becomes

$$G_{r \to u}(s) = \frac{s}{K}\frac{(2\beta\zeta + 1)|\gamma_1|s + 1}{\beta^2|\gamma_1|^2 s^2 + 2\beta\zeta|\gamma_1|s + 1} \qquad (6.21)$$

We define $\hat{s} = |\gamma_1|s$ as a scaled Laplace transform variable which allows us to rewrite Equation (6.21) as

$$G_{r \to u}(s) = \frac{\hat{s}}{K|\gamma_1|}\frac{(2\beta\zeta + 1)\hat{s} + 1}{\beta^2\hat{s}^2 + 2\beta\zeta\hat{s} + 1} \qquad (6.22)$$

The scaling with $|\gamma_1|$ in the Laplace domain naturally leads to a scaling in the time domain with $\hat{t} = \frac{t}{|\gamma_1|}$, where $\hat{t}$ represents the normalized time. The desired control signal response for a given step setpoint change of magnitude $\bar{r}$ has an initial change of $\frac{2\beta\zeta+1}{K|\gamma_1|\beta^2}\bar{r}$ and then exponentially decays to zero following a second order response with a normalized time constant β

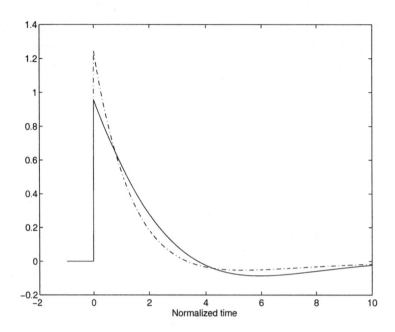

Figure 6.4: *Desired control signal trajectory with $\beta = 2$ (solid: $\zeta = 0.707$; dash-dotted: $\zeta = 1.0$)*

and a damping factor ζ. For a Type A integrating process with its corresponding values of K and $|\gamma_1|$, the choice of parameter β can be based on hard constraints on the maximum desired allowable change in the control signal and/or the desired response speed of the control signal. A smaller β corresponds to a higher performance specification (namely a faster closed-loop dynamic system response) leading to a larger initial change in the control signal. On the other hand, a larger β corresponds to a lower performance specification (i.e. a slower closed-loop dynamic system response) and a smaller initial change in the control signal response. For this type of lag dominant process, we suggest the damping factor ζ be chosen as either 1 or 0.707. The difference between these two choices for ζ in terms of their effect on the desired control signal trajectory for a unit step setpoint change is shown in Figure 6.4 with $K|\gamma_1| = 1$. The effect of β on the control signal response is illustrated in Figure 6.5 with $\zeta = 1$.

Type B: $\gamma_1 > 0$
This type of integrating process is referred to as being lead dominant. Here we select

$$\tau = \beta\gamma_1 \tag{6.23}$$

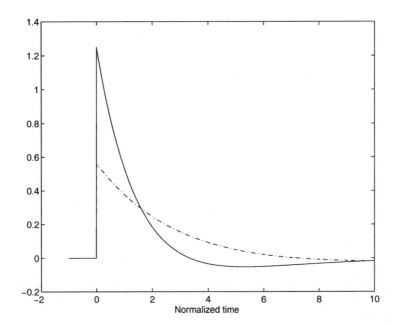

Figure 6.5: *Desired control signal trajectory with $\zeta = 1$ (solid: $\beta = 2$; dash-dotted: $\beta = 4$)*

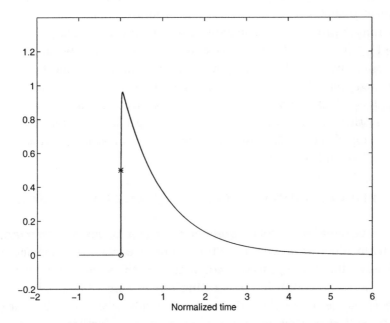

Figure 6.6: *Desired control signal trajectory with $\beta = 0.1$ ('o': initial value for $\zeta = 5$; '*': initial value for $\zeta = 5.025$)*

Then, Equation (6.18) becomes

$$G_{r \to u}(s) = \frac{s}{K} \frac{(2\beta\zeta - 1)\gamma_1 s + 1}{\beta^2 \gamma_1^2 s^2 + 2\beta\zeta\gamma_1 s + 1} \tag{6.24}$$

and its scaled form with respect to $\hat{s} = \gamma_1 s$ is given by

$$G_{r \to u}(s) = \frac{\hat{s}}{K\gamma_1} \frac{(2\beta\zeta - 1)\hat{s} + 1}{\beta^2 \hat{s}^2 + 2\beta\zeta\hat{s} + 1} \tag{6.25}$$

In order to avoid an undesirable inverse response in the control signal, the requirement is that $(2\beta\zeta - 1) \geq 0$ must be satisfied. This condition dictates that, for a smaller choice of β, a larger value for ζ has to be selected. The poles of the second order system given in Equation (6.25) are $-\frac{\zeta}{\beta}(1 \pm \sqrt{1 - \frac{1}{\zeta^2}})$. Therefore, as ζ increases, one pole becomes larger in its magnitude while the other approaches zero. Under these conditions, the control signal response tends to a first order response with time constant equal to $\frac{\beta}{\zeta - \sqrt{\zeta^2 - 1}}$.

The desired control signal response for a given step setpoint change of $\bar{r}$ has an initial change of $\frac{2\beta\zeta - 1}{K\gamma_1\beta^2}\bar{r}$. Our suggestion for choosing β and ζ is to first select β such that τ is on the same order of magnitude as the larger of the dominant process time constant or delay, and then to adjust ζ to satisfy $(2\beta\zeta - 1) \geq 0$ and to determine the closed-loop response speed, with a larger ζ corresponding to a faster response. Figure 6.6 shows the trajectories of the desired control signal for a unit step setpoint change with $\zeta = 5$ and 5.025, $\beta = 0.1$ and $K\gamma_1 = 1$. These two trajectories are almost identical except in their initial responses to the setpoint change. This difference has a significant effect on the response speed of the closed-loop system, as will be illustrated in Section 6.7.

6.3   PID PARAMETERS: LEAST SQUARES APPROACH

The control signal trajectories presented in the previous section require relatively little information about the process to be controlled. For instance, the specification for stable processes contains only the steady state process gain K and the specification for integrating processes contains only K and γ_1. However, the design of the PID controller itself, with its limited degrees of freedom, will ultimately have to be based on some further process information. The information to be used here is given by the frequency response of the process, $G(jw)$. From the desired transfer function of the control signal response to a setpoint change, we can readily find the desired closed-loop frequency response $G_{r \to y}(jw)$ as

$$G_{r \to y}(jw) = G_{r \to u}(jw)G(jw) \tag{6.26}$$

Our objective is to find the PID controller parameters such that the actual closed-loop frequency response is in some sense close to the desired closed-loop frequency response $G_{r \to y}(jw)$. However, the direct approach to this problem leads to a nonlinear optimization problem. Instead, we choose to work with the equivalent open-loop transfer function because, in this case, the problem becomes linear in the controller parameters, enabling us to consider a linear least squares approach to solving this problem.

From the desired closed-loop frequency response, the desired open-loop frequency response can be obtained from

$$G_{ol}(jw) = \frac{G_{r \to y}(jw)}{1 - G_{r \to y}(jw)} \tag{6.27}$$

The actual open-loop transfer function, $G'_{ol}(jw)$, for the process under PID control is given by

$$G'_{ol}(jw) = \frac{c_2(jw)^2 + c_1 jw + c_0}{jw} G(jw) \tag{6.28}$$

Our task is to find the coefficients c_2, c_1 and c_0 such that the sum of squared errors between the desired and actual open-loop frequency response is minimized over a set of frequencies $\{w_i\}$. We choose the loss function to be

$$V = \sum_i \left| G_{ol}(jw_i) - \frac{G(jw_i)}{jw_i} \left(c_2(jw_i)^2 + c_1 jw_i + c_0 \right) \right|^2 \tag{6.29}$$

which can be rewritten as

$$V = \sum_i |G_{ol}(jw_i) - \phi(jw_i)\theta|^2 \tag{6.30}$$

where

$$\phi(jw_i) = \left[jw_i G(jw_i) \quad G(jw_i) \quad -j\frac{G(jw_i)}{w_i} \right]$$

and

$$\theta^T = [c_2 \quad c_1 \quad c_0]$$

The minimum of this loss function V is given by applying the standard least squares result

$$\hat{\theta} = \left(\sum_i \phi^*(jw_i)\phi(jw_i) \right)^{-1} \left(\sum_i \phi^*(jw_i)G_{ol}(jw_i) \right) \tag{6.31}$$

6.3.1 Illustrative example

The following example is used to give some insight into which frequency region is better for PID controller design by comparing the results obtained using a low frequency region in the solution of Equation (6.31) with results obtained using the crossover frequency region. Consider the system given by the following transfer function (Lilja, 1990)

$$G(s) = \frac{1}{(s+1)^8} \qquad (6.32)$$

This process has an approximate settling time of 15 sec. Therefore, the lead element in the control signal trajectory specification for stable processes, $\alpha\tau$, is chosen to be $\frac{15}{5} = 3$, which is approximately equal to the dominant process time constant. Then we let $\tau = \frac{3}{\alpha}$ and tune the parameter α to determine the closed-loop response speed. The transfer function from the setpoint to the desired control signal is given by

$$G_{r \to u}(s) = \frac{\alpha\tau s + 1}{\tau s + 1} = \frac{3s + 1}{\frac{3}{\alpha}s + 1} \qquad (6.33)$$

and the desired closed-loop transfer function relating the setpoint to the process output is

$$G_{r \to y}(s) = \frac{3s + 1}{(\frac{3}{\alpha}s + 1)(s + 1)^8} \qquad (6.34)$$

The desired closed-loop process output responses for a unit step setpoint change are shown in Figure 6.7 for $\alpha = 1$ and $\alpha = 6$. With $\alpha = 1$, the desired closed-loop response is identical to the open-loop step response.

Use of the Low Frequency Region

We now concentrate only on the case of $\alpha = 6$ which corresponds to the faster closed-loop performance specification. We have specified the frequency region used in the design as $0.0001 \leq w_i \leq 0.2$ radians/sec. This region has been uniformly discretized to generate 200 frequencies. By applying the least squares algorithm in Equation (6.31), we obtain the PID controller parameters $K_c = 0.93$, $\tau_I = 5.14$ and $\tau_D = 1.80$. The actual closed-loop frequency response relating the setpoint to the process output obtained using these PID controller settings is compared with the desired closed-loop frequency response in Figure 6.8. The closed-loop frequency response fit is very good in the low frequency region, as expected. On the other hand, the fit in the higher frequency region is not as good.

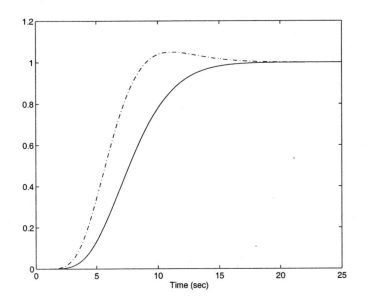

Figure 6.7: *Desired closed-loop process output responses to a unit step setpoint change (solid: $\alpha = 1$; dash-dotted: $\alpha = 6$)*

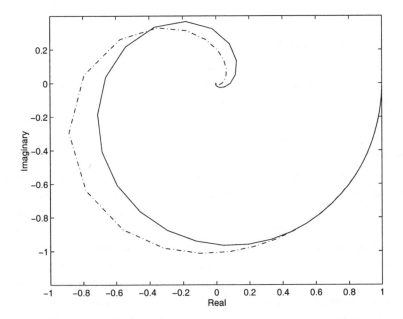

Figure 6.8: *Comparison of the desired and actual closed-loop frequency responses using low frequency region (solid: desired; dash-dotted: actual)*

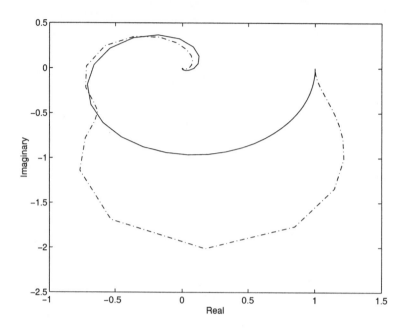

Figure 6.9: *Comparison of the desired and actual closed-loop frequency responses using crossover frequency region (solid: desired; dash-dotted: actual)*

Use of the Crossover Frequency Region

For closed-loop stability, the Nyquist plot of the actual open-loop transfer function $G'_{ol}(jw)$ must not enclose the $(-1, 0)$ point on the complex plane. This time, we have investigated the PID controller parameter solutions using frequency information around the desired closed-loop crossover frequency. With $\alpha = 6$, the crossover frequency is in the vicinity of $w = 0.6$ radians/sec. Therefore, we have chosen $0.4 \leq w_i \leq 0.8$ radians/sec, discretized using a step size of 0.001, for the design. Applying the solution in Equation (6.31), we obtain the PID controller parameters $K_c = 0.81$, $\tau_I = 2.30$ and $\tau_D = 3.34$. Note that there are quite large discrepancies between these parameters and those obtained using low frequency information. Figure 6.9 shows the desired and actual closed-loop frequency responses relating the setpoint to the process output for this new set of parameters. The comparison with Figure 6.8 shows dramatically different results, in that the closed-loop frequency response errors are significant in the low frequency region. As expected, the fit is quite good in the crossover region.

Figure 6.10 presents the closed-loop simulations comparing the two sets of PID controllers for the case $\alpha = 6$, with a unit step setpoint change at

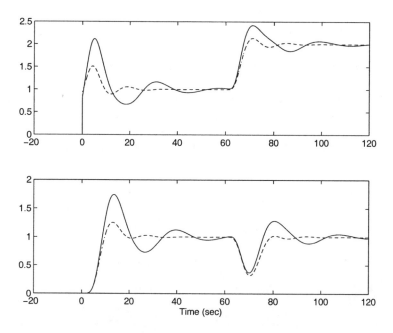

Figure 6.10: *Comparison of closed-loop responses (solid: use of crossover frequency region; dash-dotted: use of low frequency region). Upper diagram: controller output; lower diagram: process output*

$t = 0$ followed by a negative unit step load disturbance at $t = 60$ sec. We have simulated the closed-loop system with the derivative action, including a first order filter with time constant equal to $0.1\tau_D$, applied only to the measurement (see Figure 6.2). From this comparison, we can see that the PID controller designed using low frequency information performs much better than the one designed using the crossover frequency information in terms of both setpoint response and disturbance rejection. It also yields performance that is much closer to the desired response (compare Figure 6.10 with Figure 6.7). These results would seem to call into question the popular belief that the crossover frequency is the most important frequency region for PID controller design.

6.4   PID PARAMETERS: USE OF ONLY TWO FREQUENCIES

Let us begin by rewriting the loss function in Equation (6.29) in the following form

$$V = \sum_i |W(jw_i)|^2 |Y(jw_i) - [c_2(jw_i)^2 + c_1 jw_i + c_0]|^2 \qquad (6.35)$$

where $W(jw_i) = \frac{G(jw_i)}{jw_i}$ and

$$Y(jw) = \frac{G_{ol}(jw)jw}{G(jw)} \qquad (6.36)$$

$$= Y_R(w) + jY_I(w) \qquad (6.37)$$

where $Y_R(w)$ and $Y_I(w)$ are the real and imaginary parts of $Y(jw)$, respectively.

We now focus our attention on the $Y(jw)$ function. Note that from Equation (6.35) the frequency domain error between the desired open-loop frequency response $G_{ol}(jw)$ and the actual open-loop frequency response $G'_{ol}(jw) = C(jw)G(jw)$ is zero for all w if

$$Y(jw) = c_2(jw)^2 + c_1 jw + c_0 \qquad (6.38)$$

for a PID controller, or if

$$Y(jw) = c_1 jw + c_0 \qquad (6.39)$$

for a PI controller. Hence, in order to guarantee a small frequency domain error, the structure of $Y(jw)$ for a PID controller must satisfy, for $w_{min} \leq w \leq w_{max}$, and some constants β_0, β_1, and β_2,

$$Y(jw) \approx \beta_2(jw)^2 + \beta_1 jw + \beta_0 \qquad (6.40)$$

or in terms of its real and imaginary parts

$$Y_R(w) \approx \beta_0 - \beta_2 w^2 \tag{6.41}$$

and

$$Y_I(w) \approx \beta_1 w \tag{6.42}$$

Equations (6.41) and (6.42) indicate that the graph of the real part of $Y(jw)$ versus w^2 should behave like a straight line with slope $-\beta_2$ and intercept β_0, and the graph of the imaginary part of $Y(jw)$ versus w should behave like a straight line passing through the origin with slope β_1 for this error to be small.

Through the next two examples, we examine how both the real and imaginary parts of the $Y(jw)$ function behave as a function of frequency for two different processes.

Example 6.1. Consider the first order system

$$G(s) = \frac{1}{s+1} \tag{6.43}$$

We let $\alpha\tau = 1$ and therefore $\tau = \frac{1}{\alpha}$. Then the desired closed-loop transfer function from setpoint to the control signal is given by

$$G_{r \to u}(s) = \frac{s+1}{\frac{1}{\alpha}s+1} \tag{6.44}$$

and $Y(jw)$ from Equation (6.37) is given by

$$Y(jw) = \alpha jw + \alpha \tag{6.45}$$

with real part

$$Y_R(w) = \alpha \tag{6.46}$$

and imaginary part

$$Y_I(w) = \alpha w \tag{6.47}$$

Equation (6.46) represents a horizontal line with respect to w^2 that intersects the ordinate axis at α, and Equation (6.47) is a straight line with respect to w through the origin with slope α. By setting $c_1 = c_0 = \alpha$ and $c_2 = 0$ (or equivalently $K_c = c_1 = \alpha$, $\tau_I = \frac{c_1}{c_0} = 1$ and $\tau_D = \frac{c_2}{c_1} = 0$), we obtain a perfect fit between the desired open-loop frequency response $G_{ol}(jw)$ and the actual open-loop frequency response $G'_{ol}(jw) = C(jw)G(jw)$ at all frequencies, regardless of the choice of α. Therefore, any arbitrary closed-loop performance may be achieved for this first order process using a PI controller.

Example 6.2. Consider a first order plus time delay system

$$G(s) = \frac{e^{-ds}}{s+1} \qquad (6.48)$$

The closed-loop performance specification $G_{r \to u}(s)$ is chosen in the same form as Equation (6.44). However the desired closed-loop transfer function between the setpoint and the process output response is given by

$$G_{r \to y}(s) = \frac{e^{-ds}}{\frac{1}{\alpha}s+1} \qquad (6.49)$$

and therefore $Y(jw)$ is given by

$$Y(jw) = \frac{(jw+1)jw}{\frac{1}{\alpha}jw+1-e^{-djw}} \qquad (6.50)$$

$$= \frac{(jw+1)}{\frac{1}{\alpha} + \frac{1-e^{-djw}}{jw}} \qquad (6.51)$$

It is obvious by inspection of Equation (6.51) that the real and imaginary parts of Y no longer exactly satisfy Equations (6.41) and (6.42).

For instance, suppose that we choose $\alpha = 3$ for the design. Figure 6.11 shows that the real and imaginary parts of Y behave like straight lines against w^2 and w, respectively, with a small time delay ($d = 0.1$) indicating that it is possible to obtain a perfect fit between the desired open-loop frequency response and the actual open-loop frequency response at all frequencies using a PID controller. However, for a large time delay ($d = 5$), a good fit between the desired open-loop frequency response and the actual open-loop frequency response could only be obtained using a PID controller in the lower frequency region where the real and imaginary parts of Y behave like straight lines.

Proposed Solutions for PID Controller Parameters

For Example 6.1, it is clear that any two frequencies could be used to construct straight lines to fit both the real and imaginary parts of Y. However, for Example 6.2, it would appear to be better to choose two frequencies from the lower frequency region to construct approximate straight line fits. Then the slope and intercept values of these straight lines give simple solutions to the parameters c_0, c_1 and c_2. Note that, if the trajectories of the real

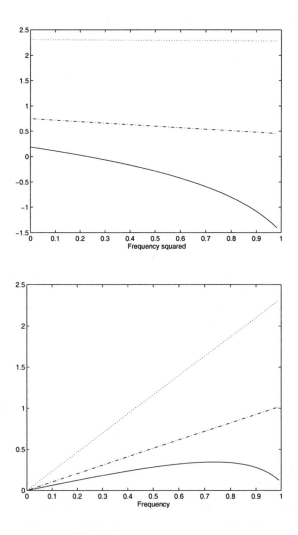

Figure 6.11: *Frequency response of Y for Example 6.2 (solid: d=5; dash-dotted: d=1; dotted: d=0.1). Upper diagram: real part of Y; lower diagram: imaginary part of Y*

and imaginary parts of Y closely satisfy Equations (6.41) and (6.42), these simple solutions will be close to the least squares solution given by Equation (6.31) within the frequency region where the two frequencies have been selected.

In summary, given the process frequency response at two frequencies $G(jw_1)$ and $G(jw_2)$ with $w_1 < w_2$, we can compute the parameters c_0, c_1 and c_2 from $Y(jw)$ as follows:

Slope of Y_R vs w^2:

$$c_2 = -\frac{Y_R(w_1) - Y_R(w_2)}{w_1^2 - w_2^2} \tag{6.52}$$

Intercept of Y_R vs w^2:

$$c_0 = Y_R(w_1) + c_2 w_1^2 \tag{6.53}$$

Slope of Y_I vs w:

$$c_1 = \frac{Y_I(w_1)}{w_1} \tag{6.54}$$

The values for these coefficients can then be substituted directly into Equations (6.4)-(6.6) to obtain the final PID controller parameters. For a PI controller, the proportional gain and integral time constant remain the same and the derivative time constant is set to zero.

6.5 CHOICE OF FREQUENCY POINTS

In this section, we decide on exactly which two frequencies to use in Equations (6.52)-(6.54) in order to solve for the PID controller parameters. Our ultimate objective is to produce a PID controller that achieves a close match between the actual and desired closed-loop performance in the time domain. Which frequencies to use for PID design has been and remains an interesting question. The well-known Ziegler-Nichols frequency response PID tuning method is based on the crossover frequency of the process. However, we have found that, although the crossover frequency is very important from a stability point of view, lower frequencies are far more important from a closed-loop performance point of view.

Our goal for PID controller design is to achieve the desired closed-loop, time domain performance with respect to both the process output variable and the control signal responses. The key is to be able to link the frequency domain controller design with the time domain performance specification. To make this link, we examine the relationship between the desired closed-loop frequency response $G_{r \to y}(jw)$ and the corresponding closed-loop response $y(t)$ to a unit step change in the setpoint $r(t)$.

We begin by assuming that the unit step response $y(t)$ is approximately equal to unity for $t \geq T_s$, where T_s is the desired closed-loop settling time. We also assume that the closed-loop system is sampled with an interval Δt. It is well known that, at the sampling instants, the continuous-time step response is equal to the discrete-time step response. This property is called step response invariance with respect to discretization. Therefore, the results developed in Chapter 5 between the frequency sampling filter model and a discrete-time step response model can be applied.

Assuming that the sampling instants correspond to $t_0 = 0$, $t_1 = \Delta t$, ..., $t_{N-1} = (N-1)\Delta t$, where $N = \frac{T_s}{\Delta t}$, we know from Equation (5.4) that the following relationship exists between the closed-loop step response and the closed-loop frequency response

$$y(t_i) \approx \sum_{l=-\frac{N-1}{2}}^{\frac{N-1}{2}} G_{r \to y}\left(j\frac{2\pi l}{T_s}\right) \frac{1}{N} \frac{1 - e^{j\frac{2\pi l}{N}(i+1)}}{1 - e^{j\frac{2\pi l}{N}}} \qquad (6.55)$$

The approximation sign accounts for the fact that we have replaced the discrete-time frequency response by its corresponding continuous-time form. Equation (6.55) indicates that the closed-loop frequency response $G_{r \to y}(jw)$ evaluated at the set of frequencies, $w = 0, \frac{2\pi}{T_s}, \ldots, \frac{\pi}{\Delta t}$ radians/time, determines the step response of the closed-loop system. In addition, from the analysis of the weighting functions in Equation (6.55) performed in Chapter 5, we found that the contributions of the frequency components to the construction of the step response decrease with increasing frequency. Therefore, in order to achieve the desired closed-loop time domain performance, the frequencies that are most important for controller design can be ranked, in descending order, starting from $w = 0$, $w = \frac{2\pi}{T_s}$, $w = \frac{4\pi}{T_s}$, and so on.

Given that the desired closed-loop transfer function will match the actual closed-loop transfer function at the zero frequency with the presence of integral action in the PID controller, we propose to use the frequencies $\frac{2\pi}{T_s}$ and $\frac{4\pi}{T_s}$ radians/time in our design. The computations for the coefficients c_0, c_1 and c_2 are summarized below, where $w_1 = \frac{2\pi}{T_s}$ and $w_2 = 2w_1$.

$$c_0 = \frac{Y_R(w_1) - Y_R(w_2)}{3} + Y_R(w_1) \qquad (6.56)$$

$$c_1 = \frac{Y_I(w_1)}{w_1} \qquad (6.57)$$

$$c_2 = \frac{Y_R(w_1) - Y_R(w_2)}{3w_1^2} \tag{6.58}$$

Summary of PID Design Method

Step 1: Specify the desired closed-loop performance through choices of α and τ for stable processes, or β and ζ for integrating processes, to form the closed-loop transfer function relating the setpoint to the desired control signal response.

Stable processes:

$$G_{r \to u}(s) = \frac{1}{K} \frac{\alpha \tau s + 1}{\tau s + 1} \tag{6.59}$$

Type A integrating processes ($\gamma_1 < 0$):

$$G_{r \to u}(s) = \frac{s}{K} \frac{(2\beta\zeta + 1)|\gamma_1|s + 1}{\beta^2 |\gamma_1|^2 s^2 + 2\beta\zeta|\gamma_1|s + 1} \tag{6.60}$$

Type B integrating processes ($\gamma_1 > 0$):

$$G_{r \to u}(s) = \frac{s}{K} \frac{(2\beta\zeta - 1)\gamma_1 s + 1}{\beta^2 \gamma_1^2 s^2 + 2\beta\zeta\gamma_1 s + 1} \tag{6.61}$$

Step 2: Choose the desired closed-loop settling time T_s in order to determine the frequencies to be used in the design. Given the process transfer function $G(s)$, T_s can be determined by simulating the step response of the desired closed-loop transfer function $G_{r \to y} = G_{r \to u}G$.

Step 3: Calculate the frequency response $G(jw)$ and $G_{r \to u}(jw)$ at $w_1 = \frac{2\pi}{T_s}$ and $w_2 = \frac{4\pi}{T_s}$ to form the desired closed-loop frequency response

$$G_{r \to y}(jw) = G_{r \to u}(jw)G(jw)$$

Step 4: Evaluate the $Y(jw)$ function at w_1 and w_2 as

$$Y(jw) = \frac{G_{ol}(jw)jw}{G(jw)} \tag{6.62}$$

where

$$G_{ol}(jw) = \frac{G_{r \to y}(jw)}{1 - G_{r \to y}(jw)} \tag{6.63}$$

Step 5: Calculate the coefficients c_0, c_1 and c_2 using Equations (6.56)-(6.58). Convert these coefficients into the final PID controller parameters using Equations (6.4)-(6.6).

6.6 ENSURING A POSITIVE INTEGRAL TIME CONSTANT

Derivative action is naturally introduced with this design method depending on the performance specification for the closed-loop response speed. A higher performance specification will likely require a PID controller while a lower performance specification will often only require a PI controller. This will be obvious to the user by looking at the sign and magnitude of the derivative time constant. When τ_D is negative, this indicates that derivative action must not be used and τ_D should be set equal to zero. The controller then reduces to PI only, using the calculated values of K_c and τ_I. When $\frac{\tau_D}{\tau_I} \leq 0.1$, it can be safely assumed that a PI controller is sufficient and again set $\tau_D = 0$.

The minimum requirement for a stable closed-loop system with a PID controller is to have a positive sign for the integral time constant. For a given process model, this requirement can be satisfied by a proper choice for the closed-loop performance parameters. (From our experience with the proposed PID design approach, the proportional gain of the controller almost always has the correct sign. Therefore, we focus our attention on ensuring the correct sign of the integral time constant.)

Stable Processes

Let us express the process transfer function $G(s)$ using a Taylor series expansion as

$$G(s) = K(1 + g_1 s + \cdots)\tag{6.64}$$

and the transfer function relating the setpoint to the desired control signal in Equation (6.11) as

$$G_{r \to u}(s) = \frac{1}{K}(1 + \tau(\alpha - 1)s + \cdots)\tag{6.65}$$

Then

$$
\begin{aligned}
G_{r \to y}(s) &= G_{r \to u}(s)G(s) \\
&= (1 + \tau(\alpha - 1)s + \cdots)(1 + g_1 s + \cdots) \\
&= 1 + (g_1 + \tau(\alpha - 1))s + \cdots
\end{aligned}\tag{6.66}
$$

and

$$
\begin{aligned}
G_{ol}(s) &= \frac{G_{r \to y}(s)}{1 - G_{r \to y}(s)} \\
&= -\frac{1 + (g_1 + \tau(\alpha - 1))s + \cdots}{(g_1 + \tau(\alpha - 1))s + \cdots}
\end{aligned}\tag{6.67}
$$

Then we obtain

$$\lim_{w \to 0} Y(jw) = \lim_{s \to 0} \frac{G_{ol}(s)s}{G(s)} = -\frac{1}{K} \frac{1}{(g_1 + \tau(\alpha - 1))} \tag{6.68}$$

Therefore, the coefficient c_0, which represents the intercept of Y_R versus w^2, is given by

$$c_0 = -\frac{1}{K} \frac{1}{(g_1 + \tau(\alpha - 1))} \tag{6.69}$$

To ensure a positive integral time constant τ_I, the parameter c_0 must be positive for a positive controller gain and negative for a negative controller gain (see Equations (6.4) and (6.5)). Assuming the controller gain has the same sign as the process gain K, Equation (6.69) leads to the following condition on the performance parameters

$$g_1 + \tau(\alpha - 1) < 0 \tag{6.70}$$

Integrating Processes

Assume that the integrating process transfer function $G(s)$ is expressed using a Taylor series expansion as

$$G(s) = \frac{K}{s}(1 + \gamma_1 s + \gamma_2 s^2 + \cdots) \tag{6.71}$$

and that the desired closed-loop transfer function from setpoint to control signal in Equation (6.18) is also expanded in terms of a Taylor series as

$$G_{r \to u}(s) = \frac{s}{K}(1 - \gamma_1 s + (2\zeta\tau\gamma_1 - \tau^2)s^2 + \cdots) \tag{6.72}$$

which gives

$$G_{r \to y}(s) = 1 + (2\zeta\tau\gamma_1 - \tau^2 - \gamma_1^2 + \gamma_2)s^2 + \cdots \tag{6.73}$$

Then

$$
\begin{aligned}
G_{ol}(s) &= \frac{G_{r \to y}(s)}{1 - G_{r \to y}(s)} \\
&= -\frac{1 + (2\zeta\tau\gamma_1 - \tau^2 - \gamma_1^2 + \gamma_2)s^2 + \cdots}{(2\zeta\tau\gamma_1 - \tau^2 - \gamma_1^2 + \gamma_2)s^2 + \cdots}
\end{aligned}
\tag{6.74}
$$

and

$$\lim_{w \to 0} Y(jw) = -\frac{1}{K} \frac{1}{2\zeta\tau\gamma_1 - \tau^2 - \gamma_1^2 + \gamma_2} \tag{6.75}$$

Therefore, the coefficient c_0, which represents the intercept of Y_R versus w^2, is given by

$$c_0 = -\frac{1}{K} \frac{1}{2\zeta\tau\gamma_1 - \tau^2 - \gamma_1^2 + \gamma_2} \tag{6.76}$$

To obtain a positive integral time constant, the control performance parameters are required to satisfy the following condition

$$2\zeta\tau\gamma_1 - \tau^2 - \gamma_1^2 + \gamma_2 < 0 \tag{6.77}$$

For both stable and integrating processes, we can expand the stable part of the transfer function, given by the general form

$$G(s) = \frac{1 + b_1 s + b_2 s^2 + \cdots}{1 + a_1 s + a_2 s^2 + \cdots} e^{-ds} \tag{6.78}$$

into its Taylor series expansion

$$G(s) = 1 + (b_1 - a_1 - d)s + (b_2 - a_2 - a_1(b_1 - a_1) - d(b_1 - a_1) + \frac{d^2}{2})s^2 + \cdots \tag{6.79}$$

From this general expression, we can easily calculate g_1 in Equation (6.64) for a stable process, and γ_1 and γ_2 in Equation (6.71) for an integrating process.

6.7   SIMULATION STUDIES

In this section, we present four simulation examples to illustrates the proposed PID controller design method and to compare it with other popular design methods found in the literature. We begin by studying two stable processes followed by two integrating processes. Both stable processes are first order plus delay systems, one with a deadtime to time constant ratio less than one (0.5) and the other with a ratio much greater than one (5.0). Both types of stable processes are frequently encountered in the process industries and therefore represent a reasonable basis for comparison.

Example 6.3. Consider the problem of PID controller design for the following plant transfer function model

$$G(s) = \frac{e^{-5s}}{10s + 1} \tag{6.80}$$

We have applied our design method using three different values of α (0.5, 1.0 and 1.5) to provide insight into the role of this key performance-related parameter. In all cases, the parameter τ has been selected according to $\tau = \frac{T}{\alpha}$,

	$\alpha = 0.5$	$\alpha = 1$	$\alpha = 1.5$	IMC	Z-N	Z-A
K_c	0.421	0.728	0.966	1	2.275	1.993
τ_I	10.52	10.93	11.20	12.5	8.59	14.83
τ_D	0.414	0.693	0.921	2	2.15	1.92

Table 6.1: *PID parameters for Example 6.3*

where $T = 10$ for this example. The desired closed-loop settling times T_s were estimated directly from the open-loop settling time of approximately 55 and the choice of α according to $T_s = \frac{55}{\alpha}$. The PID controller parameters for these three cases are summarized in Table 6.1. The closed-loop performance for each PID controlled system was evaluated through simulations. For each case, a setpoint change of magnitude 1.0 was introduced at $t = 0$ and a step load disturbance of magnitude 1 was introduced at $t = 200$, entering at the process input. The closed-loop responses are shown in Figure 6.12. It is very clear from this figure that the responses closely match the desired responses. For instance, with $\alpha = 0.5$, the initial change in the control signal for a unit setpoint change is close to 0.5 and the settling time of the control signal is approximately equal to $5 \times \tau = 5 \times 20 = 100$. Figure 6.12 shows that the speed of the process output response increases with increasing values of α as expected, but without any oscillations, even for $\alpha = 1.5$.

For comparison purposes, we have applied the design methods of Ziegler and Nichols (Z-N) (1942), Rivera *et al.* (IMC) (1986) (Case F) and Zhuang and Atherton (Z-A) (1993) to this example. The IMC method requires choice of the IMC filter time constant which we selected to be equal to 10. (This performance specification is equivalent to a choice of $\alpha = 1$ for our design.) For the Z-A design, we have used their Table 7 for setpoint changes with the derivative action, including a filter, in the feedback path. The closed-loop responses are given in Figure 6.13. For the setpoint change, the initial change in the controller output is very large for the Z-N design (368%) and for the Z-A design (272%). In addition, the control signals for these two designs exhibit significant oscillations before reaching their final steady-state values. Both of these characteristics are undesirable in a process control context due to modelling inaccuracies, saturation of final control elements, wear on final control elements, interactions with other process variables, and operator distress (Harris and Tyreus, 1987). On the other hand, the IMC design and our designs are much less aggressive. From looking at the corresponding process output responses, the Z-N and Z-A

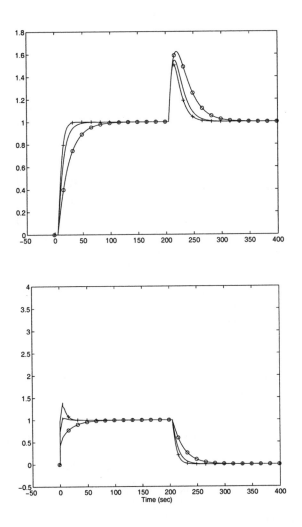

Figure 6.12: *Setpoint and load disturbance responses for Example 6.3 (solid with 'o': α = 0.5; solid: α = 1; solid with '+': α = 1.5). Upper diagram: process output; lower diagram: control signal*

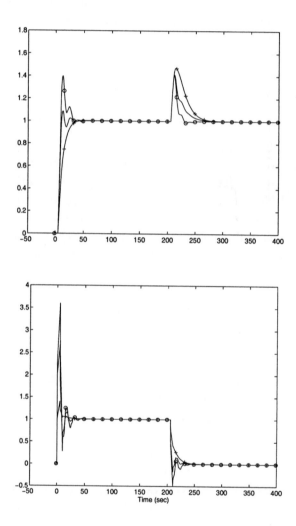

Figure 6.13: *Setpoint and load disturbance responses for Example 6.3 (solid with 'o': Z-N; solid: Z-A; solid with '+': IMC). Upper diagram: process output; lower diagram: control signal*

designs exhibit overshoot while the IMC design and our designs produce a
monotonic rise to the new setpoint value. The load responses are consistent
in that the Z-N and Z-A designs produce faster recovery times than the
IMC design and our designs due to more aggressive changes in the control
signals. Overall, the IMC design provides a level of performance which falls
in between our designs of $\alpha = 1.0$ and 1.5.

Example 6.4. Consider the problem of PID controller design for the
following plant transfer function model

$$G(s) = \frac{e^{-50s}}{10s + 1} \tag{6.81}$$

For our design, we have used a conservative choice for the parameter α of
0.25 and $\tau = \frac{T}{\alpha} = 40$. This corresponds to a desired initial change in the
control signal of 25% of its final value and an approximate settling time for
the control signal of 160. The desired settling time for the closed-loop system (T_s) was estimated from the settling times of the control signal and the
open-loop process to be 250. For this example, we are not going to compare
our results with the Ziegler-Nichols method because Åström *et al.* (1992)
recommend against using this design method for processes of this type with
delay to time constant ratios greater than unity. However, we do compare
our results with the PID tuning rules proposed by Cohen and Coon (C-C)
(1953) as suggested by Åström *et al.* (1992) to be a reasonable alternative
in this case.

We also compare our results with the IMC-PID design proposed by
Rivera *et al.* (1986) (Case F) and with the tuning suggested by Zhuang
and Atherton (Z-A) (1993). To make the comparison fair between our design and IMC, we specify the performance level to be the same in both cases,
i.e. the value for the IMC filter time constant was selected to be equal to
40. This satisfies the condition for Case F that the filter time constant be
greater than one half the process delay. PID controller parameters for these
four designs are summarized in Table 6.2.

In each simulation experiment, a setpoint change of magnitude 1.0 was
introduced at $t = 0$ and a step load disturbance of magnitude 1 entered
at $t = 500$. The closed-loop responses are compared in Figure 6.14. The
results show that both the IMC and Z-A designs produce more aggressive
responses to the setpoint and load changes than our design, and the C-C
design produces a more sluggish response. However, our design provides a
much smoother response than the other three methods in both the control
signal and the process output and we feel this is a direct result of our design

$\alpha = 0.25$	IMC	C-C	Z-A
K_c	0.275	0.538	0.517	0.508
τ_I	24.85	35.00	58.49	33.61
τ_D	3.72	7.14	9.52	13.28

Table 6.2: *PID parameters for Example 6.4*

philosophy. For instance, note that in Figure 6.14, the control signal follows almost exactly the specified trajectory for a value of $\alpha = 0.25$ and $\tau = 40$ and it is this trajectory that produces the smooth process output response.

One final point worth making is that the models in Examples 6.3 and 6.4 were used in different ways by the five design methods. In our method, we used the model to calculate the process frequency response at the two frequencies w_1 and w_2. In the IMC and Cohen-Coon methods, the transfer function model parameters were used directly in the respective formulae. For the Ziegler-Nichols and Zhuang-Atherton methods, the critical gain and period were calculated from the model.

Example 6.5. Consider the integrating process

$$G(s) = \frac{1}{s}H(s) = \frac{1}{s(s+1)^3} \tag{6.82}$$

which has been used by Åström and Hägglund (1995) to test their tuning methods. Recall that in our performance specification for integrating processes we require process information in terms of K and γ_1. For this example, $K = 1$ and $\gamma_1 = -3$, as determined by using the general expression in Equation (6.79). For this Type A (lag dominant) integrating process, the closed-loop transfer function from the setpoint to the desired control signal response is chosen to be

$$G_{r \to u}(s) = \frac{s}{K} \frac{(2\beta\zeta + 1)|\gamma_1|s + 1}{\beta^2|\gamma_1|^2 s^2 + 2\beta\zeta|\gamma_1|s + 1} \tag{6.83}$$

The parameter ζ is set to be 1 and β is used to adjust the performance, where the desired closed-loop time constant is $\beta|\gamma_1|$. Recall that a smaller β corresponds to a faster desired closed-loop system response speed. The PID controller parameters, calculated using an estimate of the desired closed-loop settling time given by $T_s = 7\beta|\gamma_1|$, are shown in Table 6.3 for three different choices for β. Closed-loop simulations are shown in Figure 6.15

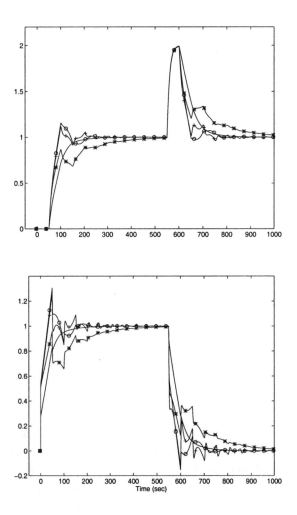

Figure 6.14: *Setpoint and load disturbance responses for Example 6.4 (solid with '+': Z-A; solid: our design; solid with 'o': IMC; solid with '*': C-C). Upper diagram: process output; lower diagram: control signal*

β	K_c	τ_I	τ_D
0.5	0.558	7.95	1.55
1	0.36	10.8	1.2
3	0.16	22.0	0.77

Table 6.3: *PID controller parameters for Example 6.5*

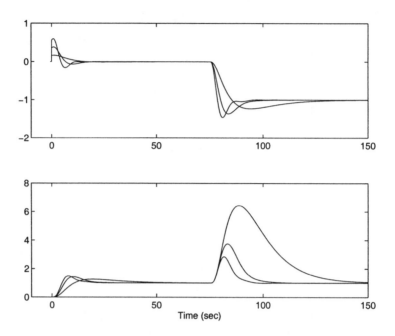

Figure 6.15: *Setpoint and load disturbance responses for Example 6.5 (fastest response speed: $\beta = 0.5$; medium response speed: $\beta = 1$; slowest response speed: $\beta = 3$). Upper diagram: control signal; lower diagram: process output*

for a unit step setpoint change at $t = 0$ and unit step load disturbance at $t = 75$. From this figure, it is clear that β provides an effective means for adjusting the closed-loop performance.

For $\beta = 0.5$, our PID controller parameters are approximately equal to the parameters obtained by Åström and Hägglund using their maximum sensitivity parameter $M_s = 2.0$ ($K_c = 0.67$, $\tau_I = 7.6$, $\tau_D = 1.7$). We have made a direct comparison of the performance of these two PID controllers in Figure 6.16, without using Åström and Hägglund's additional setpoint weighting parameter. From this figure we can see that both PID controllers give about the same setpoint response and disturbance rejection. However, some oscillations in the closed-loop responses did appear with Åström and Hägglund's settings that seemed to be avoided with our approach.

Example 6.6. The following model of an integrating process is a modified version of the transfer function presented in EnTech (1993) for the dryer steam pressure associated with paper machine dryer cans. We have added

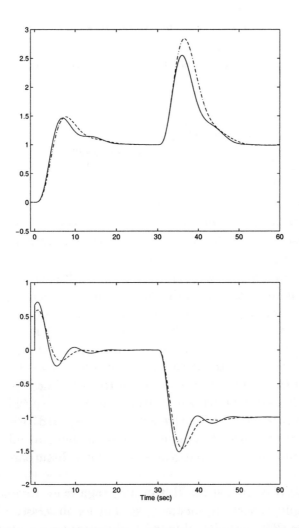

Figure 6.16: *Setpoint and load disturbance responses for Example 6.5 (solid: Åström-Hägglund (M_s = 2.0); dash-dotted: β = 0.5). Upper diagram: process output; lower diagram: control signal*

ζ	K_c	τ_I	τ_D
5	0.48	18.5	0
5.3	1.6	26.0	2.0

Table 6.4: *PID controller parameters for Example 6.6*

5 seconds of delay to this transfer function to make it more challenging.

$$G(s) = \frac{0.005(300s + 1)}{s(20s + 1)}e^{-5s} \qquad (6.84)$$

This is a Type B (lead dominant) integrating process with $K = 0.005$ and $\gamma_1 = 300 - 20 - 5 = 275$ determined using the general expression in Equation (6.79). We first apply our design method and then make some comparisons with the IMC-PID design.

First, the parameter β is chosen to be 0.1, which makes $\tau = 27.5$ on the same order of magnitude as the dominant process time constant of 20. For a slow closed-loop response, we have set the initial change in the desired control signal response to a unit step setpoint change equal to zero ($2\beta\zeta = 1$). This choice gives a value for ζ equal to 5. For a fast closed-loop response, we have set the initial change in the desired control signal response to a unit step setpoint change equal to $\frac{6}{K\gamma_1}$ which gives a value for ζ equal to 5.3. The desired closed-loop settling time T_s was approximated as $6\tau = 165$ for both cases. The PID controller parameters for both $\zeta = 5$ and $\zeta = 5.3$ are listed in Table 6.4. For $\zeta = 5$, the derivative time constant τ_D turns out to be a negative number and therefore is set equal to zero. These parameter values show that with a more aggressive control performance specification, a larger controller gain and more derivative action are required. Figure 6.17 shows the closed-loop responses for these two designs with a unit step setpoint change occurring at $t = 0$ and a unit step load disturbance occurring at $t = 400$.

We have also examined the IMC-PID design from Rivera *et al.* (1986) for this example. These authors suggest that for processes with a left half plane zero, the PID controller should be augmented with a first order lag to cancel this zero. We have designed an IMC-PID controller using their Case R which is for a model of the form

$$G(s) = K\frac{-ds + 1}{s(Ts + 1)} \qquad (6.85)$$

where, for this example, $K = 0.005$, $d = 5$, $T = 20$. We have added a first order lag with unit gain and time constant equal to 300 in series with the

ϵ	K_c	τ_I	τ_D
$\frac{400}{6}$	6.2	158.3	17.5
$\frac{100}{6}$	24.8	58.3	13.1

Table 6.5: *PID controller parameters for Example 6.6 (IMC design)*

PID controller to cancel the stable process zero and arrive at the process model structure in Equation (6.85). We based our choices for the IMC filter time constant ϵ on the approximate process output settling times found in Figure 6.17. To compare with the slow response design, we estimated the settling time to be 400 and selected $\epsilon = \frac{400}{6}$. For the fast response design we selected $\epsilon = \frac{100}{6}$. Table 6.5 summarizes the PID controller parameters for the IMC designs. Figure 6.18 shows the closed-loop responses for these two designs with the same setpoint and disturbance inputs used previously. In comparing the results, it is important to point out that the large discrepancy in the magnitude of the controller parameters results from the fact that two different process models have been used in the controller calculations, i.e. our design uses the true process model given by Equation (6.84) and the IMC design uses the process model given in Equation (6.85). It should also be noted that a series filter is required to implement the IMC design, but is not required with our design. Comparing Figure 6.17 with Figure 6.18, we can see that our design produces less overshoot to the setpoint change and faster disturbance rejection without oscillation.

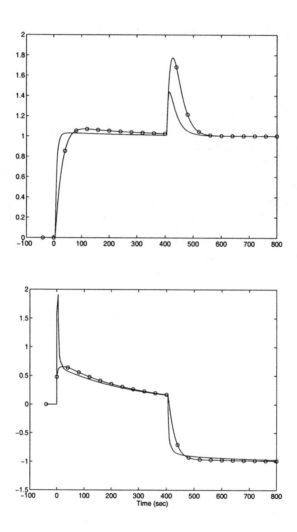

Figure 6.17: *Setpoint and load disturbance responses for Example 6.6 (solid with 'o': $\zeta = 5$; solid: $\zeta = 5.3$). Upper diagram: process output; lower diagram: control signal*

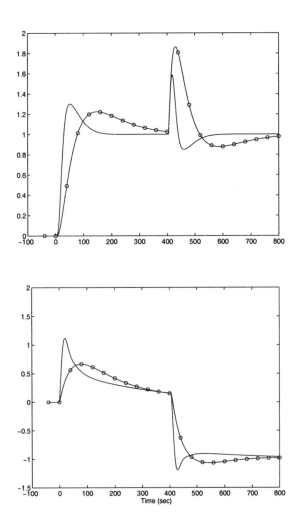

Figure 6.18: *Setpoint and load disturbance responses for Example 6.6 (solid with '0': $\epsilon = \frac{400}{6}$; solid : $\epsilon = \frac{100}{6}$). Upper diagram: process output; lower diagram: control signal*

Chapter 7

Tuning Rules for PID Controllers

7.1 INTRODUCTION

This chapter introduces new PID tuning rules, derived from the general PID design method proposed in the previous chapter, for frequently encountered first order plus delay processes and integrating plus delay processes.

This chapter consists of five sections. Section 7.2 presents the development of the PID controller tuning rules for first order plus delay processes. Sections 7.3 and 7.4 illustrate the new tuning rules using simulation and experimental studies, respectively, and compares the results with those obtained using the IMC-PID tuning rules. Section 7.5 presents the development of the PID tuning rules for integrating plus delay processes.

Portions of this chapter have been reprinted from *IEE Proceedings-Control Theory and Applications* **142**, L. Wang and W.R. Cluett, "Tuning PID controllers for integrating processes", pp. 385-392, 1997, with permission from IEE.

7.2 FIRST ORDER PLUS DELAY CASE

Assume that the process can be described by the following first order plus delay transfer function

$$G(s) = \frac{Ke^{-ds}}{Ts + 1} \tag{7.1}$$

where K is the process gain, T is the process time constant and d is the process delay. For this stable process, we choose the closed-loop transfer function from the setpoint to the desired control signal to be

$$G_{r \to u}(s) = \frac{1}{K} \frac{Ts + 1}{\frac{T}{\alpha}s + 1} \tag{7.2}$$

leading to the following desired closed-loop transfer function from the setpoint to the process output

$$
\begin{aligned}
G_{r \to y}(s) &= G_{r \to u}(s)G(s) \\
&= \frac{e^{-ds}}{\frac{T}{\alpha}s + 1}
\end{aligned}
\tag{7.3}
$$

where $\frac{T}{\alpha}$ is the desired closed-loop time constant, which we denote as τ_{cl}. In order to derive the PID tuning rules, we define $\hat{s} = ds$ as a scaled Laplace transform variable, and the ratio of time constant to delay as $L = \frac{T}{d}$. The process transfer function in Equation (7.1) can then be expressed in a normalized form as

$$G(\hat{s}) = K \frac{e^{-\hat{s}}}{L\hat{s} + 1} \tag{7.4}$$

and similarly the desired closed-loop transfer function in Equation (7.3) can be re-expressed as

$$G_{r \to y}(\hat{s}) = \frac{e^{-\hat{s}}}{\frac{L}{\alpha}\hat{s} + 1} \tag{7.5}$$

We refer to $\frac{L}{\alpha}$ as the normalized desired closed-loop time constant and denote it as $\hat{\tau}_{cl}$. Therefore, the desired closed-loop time constant and its normalized form are related by $\tau_{cl} = d\hat{\tau}_{cl}$. For a step setpoint change, we would expect the process output to take approximately $(5\hat{\tau}_{cl} + 1)d = (5\tau_{cl} + d)$ time units to reach the new setpoint value.

The ratio of the process time constant to delay, L, is a measure of the difficulty in controlling a process (Åström *et al.* 1992; Fertik, 1975). For instance, processes with a small L must have a slower desired closed-loop response speed for robustness reasons, while processes with a large L permit a faster desired closed-loop response speed. Since the parameter α reflects the desired closed-loop response speed for our design, a fixed normalized closed-loop time constant $\hat{\tau}_{cl} = \frac{L}{\alpha}$ can be used for all L to achieve the same relative performance level. For example, at a given value of $\hat{\tau}_{cl}$, a small L automatically yields a small α (slow response) and a large L yields a large

α (fast response). Therefore, $\hat{\tau}_{cl}$ forms the basis for specifying closed-loop performance in the context of PID tuning rules for this class of processes.

To derive the PID tuning rules, we write the Y function defined in Equation (6.36) in terms of the scaled variable $\hat{s}$ (or $\hat{w} = dw$)

$$Y(j\hat{w}) = \frac{\frac{G_{r\to y}(j\hat{w})}{1-G_{r\to y}(j\hat{w})}jw}{G(j\hat{w})}$$

$$= \frac{1}{Kd}\frac{(Lj\hat{w}+1)j\hat{w}}{(\hat{\tau}_{cl}j\hat{w}+1-e^{-j\hat{w}})} \tag{7.6}$$

The frequency response of $Y(j\hat{w})$ is evaluated at $\hat{w}_1 = dw_1 = d\frac{2\pi}{T_s}$ and $\hat{w}_2 = 2\hat{w}_1 = 2dw_1$, where T_s is estimated from Equation (7.5) as $(5\hat{\tau}_{cl}+1)d$. Given that the part of the denominator of $Y(j\hat{w})$ involving $\hat{w}$ is dependent only on $\hat{\tau}_{cl}$, then, for a fixed $\hat{\tau}_{cl}$, we define for the first frequency w_1

$$j\hat{\tau}_{cl}\hat{w}_1 + 1 - e^{-j\hat{w}_1} = |a_1|e^{j\phi_1} \tag{7.7}$$

and for the second frequency w_2

$$j\hat{\tau}_{cl}2\hat{w}_1 + 1 - e^{-j2\hat{w}_1} = |a_2|e^{j\phi_2} \tag{7.8}$$

Then, Equation (7.6) can be written in terms of its frequency response at $\hat{w}_1$ and $2\hat{w}_1$

$$Y(j\hat{w}_1) = \frac{1}{Kd}\frac{\hat{w}_1}{|a_1|}((sin\phi_1 - cos\phi_1\hat{w}_1 L) + j(cos\phi_1 + sin\phi_1\hat{w}_1 L)) \tag{7.9}$$

$$Y(j2\hat{w}_1) = \frac{1}{Kd}\frac{2\hat{w}_1}{|a_2|}((sin\phi_2 - 2cos\phi_2\hat{w}_1 L) + j(cos\phi_2 + 2sin\phi_2\hat{w}_1 L)) \tag{7.10}$$

We now develop new expressions for the PID controller parameters using Equations (6.56)-(6.58) and Equations (6.4)-(6.6).

Proportional gain:

$$K_c = \frac{Y_I(\hat{w}_1)}{\frac{\hat{w}_1}{d}} = \frac{\hat{K}_c}{K} \tag{7.11}$$

where

$$\hat{K}_c = k_1 + k_2 L$$

is the normalized proportional gain with k_1 and k_2 defined by

$$k_1 = \frac{cos\phi_1}{|a_1|} \tag{7.12}$$

$$k_2 = \frac{sin\phi_1}{|a_1|}\hat{w}_1 \tag{7.13}$$

Integral time constant:

$$\tau_I \; = \; \frac{K_c}{\frac{4}{3}Y_R(\hat{w}_1) - \frac{1}{3}Y_R(2\hat{w}_1)} \tag{7.14}$$

$$\; = \; d\hat{\tau}_I \tag{7.15}$$

where

$$\hat{\tau}_I = \frac{m_1 + m_2 L}{m_3 + m_4 L}$$

is the normalized integral time constant with

$$m_1 \; = \; \frac{cos\phi_1}{\hat{w}_1} \tag{7.16}$$

$$m_2 \; = \; sin\phi_1 \tag{7.17}$$

$$m_3 \; = \; \frac{2}{3}\left(2sin\phi_1 - \left|\frac{a_1}{a_2}\right|sin\phi_2\right) \tag{7.18}$$

$$m_4 \; = \; \frac{4}{3}\hat{w}_1\left(\left|\frac{a_1}{a_2}\right|cos\phi_2 - cos\phi_1\right) \tag{7.19}$$

Derivative time constant:

$$\tau_D \; = \; \frac{1}{K_c}\frac{Y_R(\hat{w}_1) - Y_R(2\hat{w}_1)}{3\frac{\hat{w}_1^2}{d^2}} \tag{7.20}$$

$$\; = \; d\hat{\tau}_D \tag{7.21}$$

where

$$\hat{\tau}_D = \frac{n_1 + n_2 L}{n_3 + n_4 L}$$

is the normalized derivative time constant with

$$n_1 \; = \; sin\phi_1 - 2\left|\frac{a_1}{a_2}\right|sin\phi_2 \tag{7.22}$$

$$n_2 \; = \; \hat{w}_1\left(4\left|\frac{a_1}{a_2}\right|cos\phi_2 - cos\phi_1\right) \tag{7.23}$$

$$n_3 \; = \; 3\hat{w}_1 cos\phi_1 \tag{7.24}$$

$$n_4 \; = \; 3\hat{w}_1^2 sin\phi_1 \tag{7.25}$$

Note that the normalized PID controller parameters $\hat{K}_c$, $\hat{\tau}_I$ and $\hat{\tau}_D$ are only dependent on the ratio of the time constant to delay, L, and the normalized desired closed-loop time constant, $\hat{\tau}_{cl}$.

We can also express the actual open-loop transfer function in terms of the normalized process transfer function in Equation (7.4) and the normalized PID controller parameters

$$G_{ol}(s) = K_c(1 + \frac{1}{\tau_I s} + \tau_D s)G(s) \qquad (7.26)$$

$$= \hat{K}_c(1 + \frac{1}{\hat{\tau}_I \hat{s}} + \hat{\tau}_D \hat{s})\frac{e^{-\hat{s}}}{L\hat{s} + 1} \qquad (7.27)$$

It is clear from this expression that the open-loop transfer function is only a function of the ratio of the process time constant to delay, L, and the choice of $\hat{\tau}_{cl}$, and is independent of the actual process parameters. Therefore, for a given choice of $\hat{\tau}_{cl}$, we are able to obtain the gain margin (GM) and phase margin (PM) of the designed system with respect to L only. Since the critical frequencies associated with Equation (7.27) are with respect to $\hat{s}$, the actual crossover frequency w_c is scaled by the process delay as $w_c = \frac{\hat{w}_c}{d}$ and the critical phase margin frequency is $w_p = \frac{\hat{w}_p}{d}$.

In addition to gain and phase margins, another important measure of robustness for the designed PID control system is the allowable time delay variation, i.e. the increase in the process delay that would bring the closed-loop system to the stability boundary. The closed-loop system reaches this critical stability boundary when the change in the delay, Δd, multiplied by the critical frequency w_p equals the phase margin (in radians). That is

$$\Delta d \times w_p = PM \qquad (7.28)$$

which gives the allowable relative time delay variation (or relative delay margin)

$$RDM = \frac{\Delta d}{d} = \frac{PM}{\hat{w}_p} \qquad (7.29)$$

As with the gain and phase margins, the relative delay margin depends only on L and the choice of $\hat{\tau}_{cl}$.

Presentation of the Tuning Rules

Although Equations (7.11)-(7.25) give analytical solutions for the PID controller parameters, it would be even more convenient for the user to have tuning rules requiring a minimum number of calculations. In order to present a range of desired closed-loop response speeds, we have chosen six different values for the normalized closed-loop time constant $\hat{\tau}_{cl} = 4, 2, 1.33, 1, 0.8$ and 0.67. The corresponding value of the parameter α can be simply calculated as $\alpha = \frac{L}{\hat{\tau}_{cl}}$ and the actual desired closed-loop time constant is given by

$\tau_{cl} = d\hat{\tau}_{cl}$. In this simplified form, the tuning rules will be on a comparable level with the Ziegler-Nichols and Cohen-Coon rules in terms of ease of use, with the added benefit of having several different performance levels for the user to choose from along with their respective stability margins.

Table 7.1 lists the normalized PID controller parameters for each $\hat{\tau}_{cl}$ and the respective ranges of the stability margins for both PID and PI design (the settings for PI control are obtained from the same set of equations for PID with $\tau_D = 0$). These stability margins were calculated for $0.1 \leq L \leq 100$. Figure 7.1 shows the corresponding gain margins and phase margins, and Figure 7.3 shows the allowable relative time delay variation when a PID controller is used. Figure 7.2 shows the gain and phase margins, and Figure 7.4 shows the allowable relative time delay variation with a PI controller. All of these figures only show the values of the margins for $0.1 \leq L < 20$ because the margins remain unchanged for $L > 20$. Comparing Figure 7.1 with Figure 7.2 and Figure 7.3 with Figure 7.4, we can see that, with a lower performance specification ($4 \leq \hat{\tau}_{cl} < 1.33$), there are not significant differences between the PI and PID stability margins. However, the differences become significant with higher performance specifications ($1 \leq \hat{\tau}_{cl} \leq 0.67$). This suggests that for a lower performance specification, a PI controller is adequate while for a higher performance specification, a PID controller can provide larger stability margins and improved performance over PI. Another point worth noting is that when L is very small, the differences between the stability margins for PI and PID are insignificant regardless of the performance level. This means that a PI controller is sufficient for delay dominant systems.

To apply the new tuning rules, the user must first identify the process dynamics in terms of K, T and d. Then, a value for the normalized desired closed-loop time constant $\hat{\tau}_{cl}$ must be selected. This choice of $\hat{\tau}_{cl}$ would be dictated by the desired response speed of the closed-loop system (i.e. $T_s = (5\hat{\tau}_{cl} + 1)d$) with due consideration for the acceptable stability margins. This information is available for both PI and PID settings in Table 7.1 and Figures 7.1-7.4. With the selected value for $\hat{\tau}_{cl}$ and the value identified for $L = \frac{T}{d}$, the normalized controller parameters ($\hat{K}_c$, $\hat{\tau}_I$, $\hat{\tau}_D$) are calculated from Table 7.1. The final step is to evaluate the actual controller parameters for implementation according to Equations (7.11), (7.15) and (7.21), i.e. $K_c = \frac{\hat{K}_c}{K}$; $\tau_I = d\hat{\tau}_I$; $\tau_D = d\hat{\tau}_D$.

$\hat{\tau}_{cl}$	4	2
$\hat{K}_c$	$1.9952 \times 10^{-2} + 2.0042 \times 10^{-1}L$	$5.5548 \times 10^{-2} + 3.3639 \times 10^{-1}L$
$\hat{\tau}_I$	$\frac{9.9508 \times 10^{-2} + 9.9956 \times 10^{-1}L}{9.9747 \times 10^{-1} - 8.7425 \times 10^{-5}L}$	$\frac{1.6440 \times 10^{-1} + 9.9558 \times 10^{-1}L}{9.8607 \times 10^{-1} - 1.5032 \times 10^{-4}L}$
$\hat{\tau}_D$	$\frac{-6.9905 \times 10^{-3} + 2.9480 \times 10^{-2}L}{2.9773 \times 10^{-2} + 2.9907 \times 10^{-1}L}$	$\frac{-1.6651 \times 10^{-2} + 9.3641 \times 10^{-2}L}{9.3905 \times 10^{-2} + 5.6867 \times 10^{-1}L}$
GM_1	$7.97 - 8.56$	$4.84 - 5.31$
PM_1	$79.67 - 79.70$	$73.9 - 74.14$
RDM_1	$6.94 - 6.95$	$3.83 - 3.86$
GM_2	$7.50 - 8.28$	$4.35 - 5.02$
PM_2	$78.54 - 79.68$	$70.73 - 74.04$
RDM_2	$6.8 - 6.95$	$3.62 - 3.86$

$\hat{\tau}_{cl}$	1.33	1
$\hat{K}_c$	$9.2654 \times 10^{-2} + 4.3620 \times 10^{-1}L$	$1.2786 \times 10^{-1} + 5.1235 \times 10^{-1}L$
$\hat{\tau}_I$	$\frac{2.0926 \times 10^{-1} + 9.8518 \times 10^{-1}L}{9.6515 \times 10^{-1} + 4.2550 \times 10^{-3}L}$	$\frac{2.4145 \times 10^{-1} + 9.6751 \times 10^{-1}L}{9.3566 \times 10^{-1} + 2.2988 \times 10^{-2}L}$
$\hat{\tau}_D$	$\frac{-2.4442 \times 10^{-2} + 1.7669 \times 10^{-1}L}{1.7150 \times 10^{-1} + 8.0740 \times 10^{-1}L}$	$\frac{-3.0407 \times 10^{-2} + 2.7480 \times 10^{-1}L}{2.5285 \times 10^{-1} + 1.0132L}$
GM_1	$3.81 - 4.16$	$3.30 - 3.56$
PM_1	$69.84 - 70.70$	$65.86 - 68.50$
RDM_1	$2.78 - 2.85$	$2.23 - 2.37$
GM_2	$3.29 - 3.89$	$2.74 - 3.30$
PM_2	$64.4 - 70.64$	$57.83 - 68.50$
RDM_2	$2.5 - 2.85$	$1.89 - 2.37$

$\hat{\tau}_{cl}$	0.8	0.67
$\hat{K}_c$	$1.6051 \times 10^{-1} + 5.7109 \times 10^{-1}L$	$1.9067 \times 10^{-1} + 6.1593 \times 10^{-1}L$
$\hat{\tau}_I$	$\frac{2.6502 \times 10^{-1} + 9.4291 \times 10^{-1}L}{8.9868 \times 10^{-1} + 6.9355 \times 10^{-2}L}$	$\frac{2.8242 \times 10^{-1} + 9.1231 \times 10^{-1}L}{8.5491 \times 10^{-1} + 1.5937 \times 10^{-1}L}$
$\hat{\tau}_D$	$\frac{-3.5204 \times 10^{-2} + 3.8823 \times 10^{-1}L}{3.3303 \times 10^{-1} + 1.1849L}$	$\frac{-3.9589 \times 10^{-2} + 5.1941 \times 10^{-1}L}{4.0950 \times 10^{-1} + 1.3228L}$
GM_1	$3.00 - 3.19$	$2.80 - 2.95$
PM_1	$60.60 - 67.21$	$52.35 - 66.45$
RDM_1	$1.82 - 2.09$	$1.44 - 1.92$
GM_2	$2.39 - 2.93$	$2.11 - 2.68$
PM_2	$49.64 - 67.13$	$38.14 - 66.25$
RDM_2	$1.42 - 2.08$	$0.99 - 1.90$

Table 7.1: *Normalized PID controller tuning rules (subscript 1: PID margins; subscript 2: PI margins)*

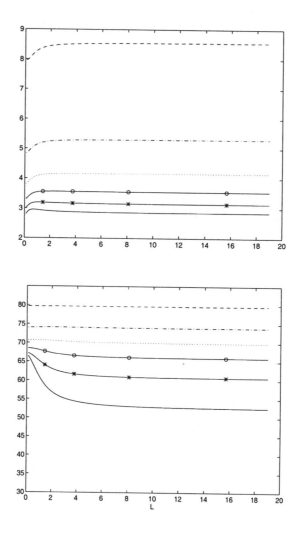

Figure 7.1: *PID stability margins as a function of L (dashed: $\hat{\tau}_{cl} = 4$; dash-dotted: $\hat{\tau}_{cl} = 2$; dotted: $\hat{\tau}_{cl} = 1.33$; solid with 'o': $\hat{\tau}_{cl} = 1$; solid with '*': $\hat{\tau}_{cl} = 0.8$; solid: $\hat{\tau}_{cl} = 0.67$). Upper diagram: gain margins; lower diagram: phase margins*

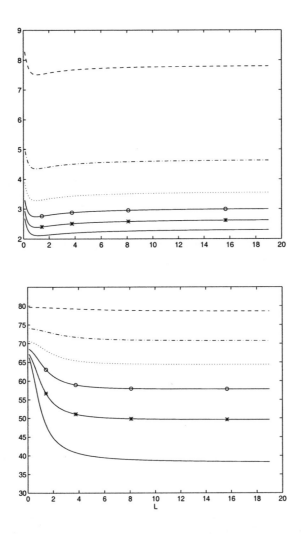

Figure 7.2: *PI stability margins as a function of L (dashed: $\hat{\tau}_{cl} = 4$; dash-dotted: $\hat{\tau}_{cl} = 2$; dotted: $\hat{\tau}_{cl} = 1.33$; solid with 'o': $\hat{\tau}_{cl} = 1$; solid with '*': $\hat{\tau}_{cl} = 0.8$; solid: $\hat{\tau}_{cl} = 0.67$). Upper diagram: gain margins; lower diagram: phase margins*

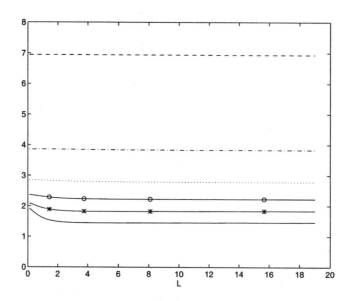

Figure 7.3: *PID relative delay margins as a function of L (dashed: $\hat{\tau}_{cl} = 4$; dashdotted: $\hat{\tau}_{cl} = 2$; dotted: $\hat{\tau}_{cl} = 1.33$; solid with 'o': $\hat{\tau}_{cl} = 1$; solid with '*': $\hat{\tau}_{cl} = 0.8$; solid: $\hat{\tau}_{cl} = 0.67$)*

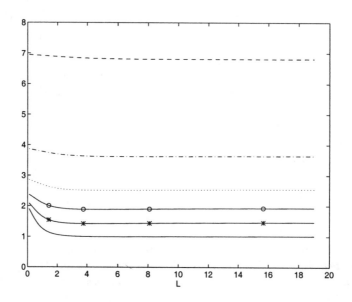

Figure 7.4: *PI relative delay margins as a function of L (dashed: $\hat{\tau}_{cl} = 4$; dashdotted: $\hat{\tau}_{cl} = 2$; dotted: $\hat{\tau}_{cl} = 1.33$; solid with 'o': $\hat{\tau}_{cl} = 1$; solid with '*': $\hat{\tau}_{cl} = 0.8$; solid: $\hat{\tau}_{cl} = 0.67$)*

7.3   EVALUATION OF THE NEW TUNING RULES: SIMULATION RESULTS

Here, we illustrate the performance of the new tuning rules using the following simulated example (Process A)

$$G(s) = \frac{-e^{-5s}}{5s + 1} \qquad (7.30)$$

Both PI and PID controllers have been designed for this process for three values of $\hat{\tau}_{cl} = 2$, 1, and 0.67, which correspond in this case to desired closed-loop time constants τ_{cl} of 10, 5 and 3.3, respectively. Table 7.2 presents the actual controller parameters for these choices of $\hat{\tau}_{cl}$. Figure 7.5 shows the closed-loop responses with PI control and Figure 7.6 shows the closed-loop responses under PID control, for a unit step setpoint change followed by a negative unit step load disturbance. The closed-loop system has been simulated with the derivative action, including a first order filter with time constant equal to $0.1\tau_D$, applied only to the measurement.

These results illustrate that with the lower performance specification ($\hat{\tau}_{cl} = 2$), there is little difference between the PI and PID performance. However, for the higher performance specifications ($\hat{\tau}_{cl} = 1$ and 0.67), the PID performance is superior in that it produces less oscillatory control signal and process output responses.

Comparisons with various existing PID tuning rules were carried out in Chapter 6 for a similar example and it was found that our design method consistently produced smoother control signal and process output responses, both of which are generally considered to be desirable in a process control application. In this section, we focus on highlighting some other distinctive features of the new tuning rules and we use the popular IMC-PID design rules of Rivera *et al.* (1986) for comparative purposes.

$\hat{\tau}_{cl}$	K_c	τ_I	τ_D
2	−0.39	5.9	0.6
1	−0.64	6.3	0.95
0.67	−0.81	5.9	1.4

Table 7.2: *PID controller parameters for Process A using new tuning rules*

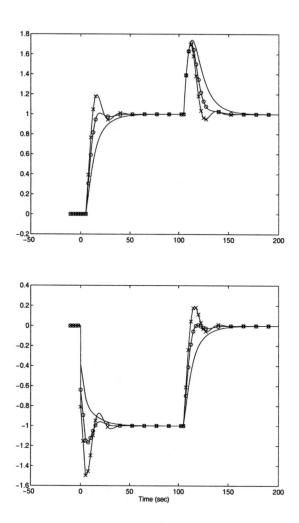

Figure 7.5: *Setpoint and load disturbance responses for Process A under PI control (solid: $\hat{\tau}_{cl} = 2$; solid with 'o': $\hat{\tau}_{cl} = 1$; solid with '×': $\hat{\tau}_{cl} = 0.67$). Upper diagram: process output; lower diagram: control signal*

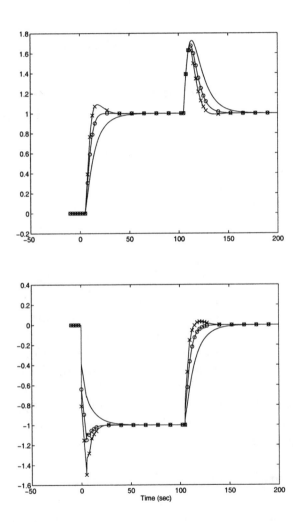

Figure 7.6: *Setpoint and load disturbance responses for Process A under PID control (solid: $\hat{\tau}_{cl} = 2$; solid with 'o': $\hat{\tau}_{cl} = 1$; solid with 'x': $\hat{\tau}_{cl} = 0.67$). Upper diagram: process output; lower diagram: control signal*

$\hat{\tau}_{cl}$	K_c	τ_I	τ_D
2	−0.6	7.5	1.67
1	−1.0	7.5	1.67
0.67	−1.3	7.5	1.67

Table 7.3: *IMC-PID controller parameters for Process A*

Desired Performance Determines Required P, I and D Actions

As can be seen from Table 7.1, all three of the normalized controller parameters vary with the choice of $\hat{\tau}_{cl}$. For Process A, Table 7.2 presents the actual controller parameters for the three different performance levels. For this example, the controller gain increases as $\hat{\tau}_{cl}$ decreases, the integral time constant increases and then decreases as $\hat{\tau}_{cl}$ decreases, and the derivative time constant increases as $\hat{\tau}_{cl}$ decreases. In addition, the amount of derivative action relative to the integral action changes as a function of $\hat{\tau}_{cl}$. For instance, at $\hat{\tau}_{cl} = 2$, $\frac{\tau_D}{\tau_I} = 0.1$ and at $\hat{\tau}_{cl} = 0.67$, $\frac{\tau_D}{\tau_I} = 0.24$.

The IMC-PID controller settings for this process (Case F in Rivera *et al.* (1986)) are given in Table 7.3 for the same performance levels (IMC filter time constant $\epsilon = \tau_{cl}$). From this table, it can be seen that only the controller gain K_c changes with $\hat{\tau}_{cl}$, and τ_I and τ_D remain unchanged with $\frac{\tau_D}{\tau_I} = 0.22$. In addition, it is actually the choice of approximation for the time delay term with the IMC design (e.g. first order Taylor series or first order Pade) that determines whether a PI or PID controller is selected (Case D versus Case F in Rivera *et al.* (1986)). With the new tuning rules, no approximations to the delay term are required and the appropriate amounts of all three controller actions are directly determined from the closed-loop performance specification.

Smooth Transition from PID to PI Performance as Filtering is Added

When noise is present in the measured process output, any derivative action should be accompanied by a filter. However, the problem is often that, as the time constant of the derivative filter is increased to cope with the measurement noise, the closed-loop performance degrades, requiring re-tuning of the controller parameters. This problem is illustrated using the IMC-PID rules applied to Process A with $\hat{\tau}_{cl} = 0.67$. Figure 7.7 shows the closed-loop responses to a negative unit step load disturbance with a derivative filter time constant equal to $0.1\tau_D$. Figure 7.8 shows the closed-loop responses

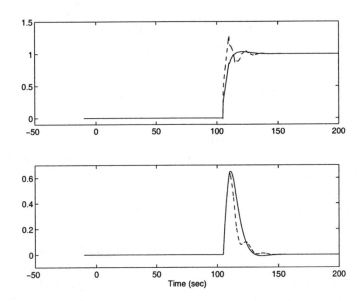

Figure 7.7: *PID control with derivative filter time constant of* $0.1\tau_D$ *for Process A (solid: new rules; dashed: IMC). Upper diagram: control signal; lower diagram: process output*

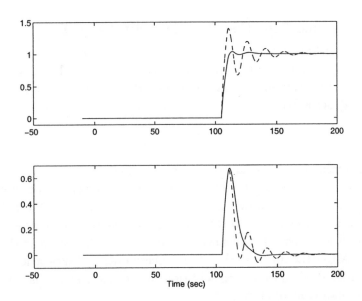

Figure 7.8: *PID control with derivative filter time constant of* $1.0\tau_D$ *for Process A (solid: new rules; dashed: IMC). Upper diagram: control signal; lower diagram: process output*

	K_c	τ_I	τ_D
IMC-PID	-17.52	5.125	0.122
New Rules	-12.51	1.15	0.1

Table 7.4: *PID controller parameters for Process B*

to the same load disturbance with a derivative filter time constant equal to $1.0\tau_D$. Clearly, the response goes from being only slightly oscillatory with little filtering to very oscillatory with more filtering. With IMC-PID tuning, the IMC filter time constant would have to be increased in order to reduce these oscillations. For comparison, Figures 7.7 and 7.8 also show the closed-loop responses under the same conditions using the new tuning rules with $\hat{\tau}_{cl} = 0.67$, where there is only a slight deterioration in performance due to the increased derivative filtering action. In fact, as the amount of filtering is increased, the performance of the PID controller gradually approaches the PI controller performance (compare Figure 7.5 for PI with Figure 7.6 for PID). This is a very desirable feature of the new tuning rules because re-tuning of the controller parameters is not necessary, regardless of the amount of filtering used.

Fast Disturbance Rejection Provided with Large L

The slow disturbance rejection properties of PID design methods based on pole-zero cancellation has been pointed out by Chien and Fruehauf (1990). To illustrate this problem, we have modified Process A to have the following dynamics (Process B)

$$G(s) = \frac{-e^{-0.25s}}{5s + 1} \tag{7.31}$$

which has a relatively large L value of 20. For fast disturbance rejection, we have chosen $\hat{\tau}_{cl} = 0.67$. PID controller parameters obtained for this process using both IMC-PID tuning rules and the new rules are shown in Table 7.4. The resulting closed-loop responses are compared in Figures 7.9, for a negative unit step load disturbance. From Figure 7.9, it is clear that, when using the IMC-PID tuning rules, the process output takes a long time to return to its setpoint value. In addition, the control signal is not smooth due to the high controller gain K_c.

To circumvent this problem of slow disturbance rejection, Chien and Fruehauf (1990) have suggested that a first order plus delay process with a large L be remodelled as an integrator plus delay process. However, with the

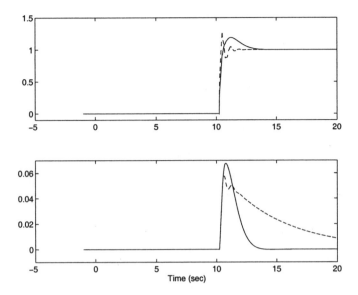

Figure 7.9: *PID Control for Process B (solid: new rules; dashed: IMC). Upper diagram: control signal; lower diagram: process output*

new tuning rules, fast disturbance rejection is achieved simply by choosing a small value for $\hat{\tau}_{cl}$ without the need for any remodelling of the process dynamics. Using the new rules, the process output is seen in Figure 7.9 to return very quickly to its setpoint value with a very smooth control signal response.

Comparing the controller parameters in Table 7.4, the IMC-PID design rules produce both a larger proportional gain and a larger integral time constant as compared to the new rules. However, it is primarily the smaller integral time constant associated with the new rules that produces the fast disturbance rejection.

7.4 EXPERIMENTS WITH A STIRRED TANK HEATER

This section presents experimental results that compare our new tuning rules and the IMC tuning rules. The apparatus used to carry out all of the experiments is a pilot-scale, continuous stirred tank water heater (see Figure 7.10 for a process schematic). Hot and cold water streams are combined to produce the feed stream. The water in the tank is heated by a steam coil and the steam line is equipped with a control valve. The water is removed from the tank using a pump. The tank is also equipped with a level sensor and

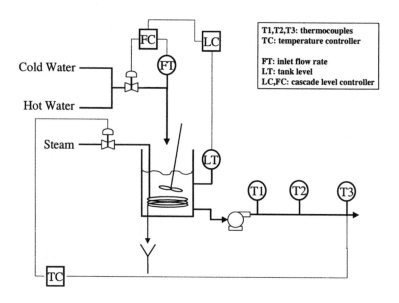

Figure 7.10: *Schematic of the stirred tank heater apparatus*

an overflow line. All experiments were performed with the tank operating
at overflow and the inlet feed water valve set to a fixed position, except in
the disturbance rejection experiments where a flow controller was used to
maintain constant feed water flow when the hot water was turned off (the
cascade level controller was not used in any of these experiments). The
steady state relationship between the steam control valve and the outlet
water temperature is nonlinear. During the experiments, the steam valve
position was operated in an approximate linear region between $0-40\%$ open.

Simple first order plus delay models for this process were derived from
step response tests. For controlling the temperature of the outlet water
stream, the manipulated input variable is the steam valve position and the
process output variable is the outlet water temperature, measured at one of
three thermocouples located at different distances from the tank outlet. The
tests involved introducing a step change in steam valve position, with the
temperature controller in manual, and observing the response in the outlet
water temperature at each of the three thermocouples. Simple graphical
methods were used to calculate the parameters of the first order plus delay
models.

The models are, for thermocouple 1

$$G_1(s) = \frac{0.4e^{-8s}}{116s + 1} \tag{7.32}$$

for thermocouple 2

$$G_2(s) = \frac{0.4e^{-40s}}{116s + 1} \tag{7.33}$$

and for thermocouple 3

$$G_3(s) = \frac{0.4e^{-80s}}{116s + 1} \tag{7.34}$$

with the time constant and time delays in seconds. Clearly, the only difference between the three models is in their time delays. PID controllers using both IMC tuning rules and the new tuning rules were calculated using the model for thermocouple 2. In order to achieve a fair comparison, the desired closed-loop time constant in both cases was set equal to 0.67×40 ($\hat{\tau}_{cl} = 0.67$).

Experiment 1: Disturbance Rejection

In the first experiment, a step load disturbance was introduced to the process in order to observe which controller settings would provide better disturbance rejection. This disturbance was created by switching off the hot feed water stream causing a sudden drop in the inlet temperature from approximately 17^oC to 5^oC. Figure 7.11 shows the experimental results, where we can see that the PID controller designed using the new rules brings the outlet water temperature back to setpoint faster than the IMC-PID settings, although the response with the new rules exhibits a larger initial deviation from the setpoint.

Experiment 2: Robustness Comparison

For the second experiment, the PID controllers were designed based on the model for thermocouple 2, but the controller itself was configured to measure the outlet water temperature from thermocouple 3. This represents an effective doubling of the process time delay from 40 to 80 sec. Figure 7.12 shows the results with the two controllers, where we have introduced a step setpoint change from 11^oC to 15^oC. We can see from this figure that the closed-loop system using the IMC-PID tuning rules is closer to being unstable and the new rules provide more robust performance.

Experiment 3: Increasing Derivative Filtering

In the third experiment, we designed and implemented both PID controllers from thermocouple 2 and then increased the amount of derivative filtering by increasing the derivative filter time constant from $0.1\tau_D$ (used in Figure 7.11) to $1.0\tau_D$ shown in Figure 7.13, while introducing the same step load disturbance used in Experiment 1. We can see by comparing Figures 7.11 and 7.13 that increasing the amount of derivative filtering yielded a somewhat more oscillatory temperature response and control signal when using the IMC-PID design as compared to the new tuning rules.

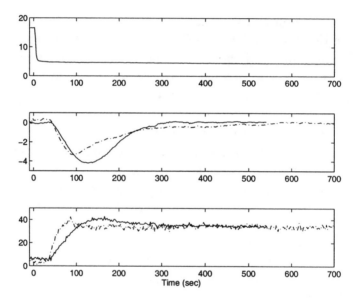

Figure 7.11: *Results from Experiment 1 (solid: new rules; dash-dotted: IMC-PID rules). Upper diagram: inlet water temperature (°C); middle diagram: deviation outlet water temperature (°C); lower diagram: control signal (% steam valve position)*

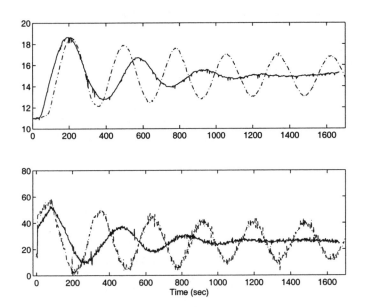

Figure 7.12: *Results from Experiment 2 (solid: new rules; dash-dotted: IMC-PID rules). Upper diagram: outlet water temperature (°C); lower diagram: control signal (% steam valve position)*

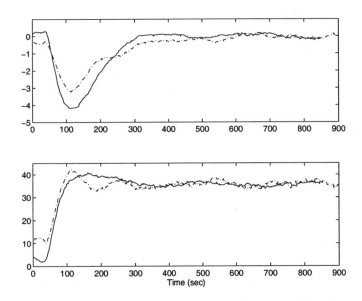

Figure 7.13: *Results from Experiment 3 (solid: new rules; dash-dotted: IMC-PID rules). Upper diagram: deviation outlet water temperature (°C); lower diagram: control signal (% steam valve position)*

7.5   INTEGRATING PLUS DELAY CASE

Assume that the process has the following transfer function

$$G(s) = \frac{K}{s} e^{-ds} \tag{7.35}$$

For this process, $\gamma_1 = -d < 0$, and therefore this is a Type A integrating process. The desired control signal specification is chosen as

$$G_{r \to u}(s) = \frac{s}{K} \frac{(2\beta\zeta + 1)ds + 1}{\beta^2 d^2 s^2 + 2\beta\zeta ds + 1} \tag{7.36}$$

By letting $\hat{s} = ds$, $G_{r \to u}(s)$ is scaled in the Laplace domain as

$$G_{r \to u}(s) = \frac{\hat{s}}{Kd} \frac{(2\beta\zeta + 1)\hat{s} + 1}{\beta^2 \hat{s}^2 + 2\beta\zeta\hat{s} + 1} \tag{7.37}$$

and the desired closed-loop transfer function is given by

$$G_{r \to y}(s) = \frac{(2\beta\zeta + 1)\hat{s} + 1}{\beta^2 \hat{s}^2 + 2\beta\zeta\hat{s} + 1} e^{-\hat{s}} \tag{7.38}$$

From the desired closed-loop transfer function, the corresponding desired open-loop transfer function is obtained as

$$
\begin{aligned}
G_{ol}(s) &= \frac{G_{r \to y}(s)}{1 - G_{r \to y}(s)} \\[2mm]
&= \frac{(2\beta\zeta + 1)\hat{s} + 1}{\beta^2 \hat{s}^2 + 2\beta\zeta\hat{s} + 1 - [(2\beta\zeta + 1)\hat{s} + 1]e^{-\hat{s}}} e^{-\hat{s}}
\end{aligned}
\tag{7.39}
$$

This then gives

$$
\begin{aligned}
Y(s) &= \frac{G_{ol}(s)s}{G(s)} \\[2mm]
&= \frac{\hat{s}^2}{Kd^2} \frac{(2\zeta\beta + 1)\hat{s} + 1}{\beta^2 \hat{s}^2 + 2\beta\zeta\hat{s} + 1 - [(2\beta\zeta + 1)\hat{s} + 1]e^{-\hat{s}}}
\end{aligned}
\tag{7.40}
$$

The two frequencies used in the controller parameter solutions are $w_1 = \frac{2\pi}{T_s}$ and $w_2 = \frac{4\pi}{T_s}$, where T_s is the desired closed-loop settling time, estimated here as $(6\beta+1)d$. The corresponding normalized frequencies are $\hat{w}_1 = dw_1 = \frac{2\pi}{6\beta+1}$ and $\hat{w}_2 = 2\hat{w}_1$. We now define

$$\hat{Y}(j\hat{w}) = (j\hat{w})^2 \frac{(2\zeta\beta + 1)j\hat{w} + 1}{\beta^2(j\hat{w})^2 + 2\beta\zeta j\hat{w} + 1 - [(2\beta\zeta + 1)j\hat{w} + 1]e^{-j\hat{w}}} \tag{7.41}$$

where
$$\hat{Y}(j\hat{w}) = Kd^2Y(j\hat{w}) \tag{7.42}$$

With our choice of $\hat{w}_1$ and $\hat{w}_2$, $\hat{Y}$ is dependent only on the performance parameters β and ζ, and is independent of the actual process parameters. We now derive new expressions for the PID controller parameters using Equations (6.56)-(6.58) and Equations (6.4)-(6.6).

Proportional gain:
$$K_c = \frac{\hat{K}_c}{dK} \tag{7.43}$$

where
$$\hat{K}_c = \frac{\hat{Y}_I(\hat{w}_1)}{\hat{w}_1}$$

Integral time constant:
$$\tau_I = d\hat{\tau}_I \tag{7.44}$$

where
$$\hat{\tau}_I = \frac{\hat{K}_c}{\hat{Y}_R(\hat{w}_1) + \frac{\hat{Y}_R(\hat{w}_1) - \hat{Y}_R(\hat{w}_2)}{3}}$$

Derivative time constant:
$$\tau_D = d\hat{\tau}_D \tag{7.45}$$

where
$$\hat{\tau}_D = \frac{\hat{Y}_R(\hat{w}_1) - \hat{Y}_R(\hat{w}_2)}{3\hat{K}_c\hat{w}_1^2}$$

Note that the normalized PID controller parameters $\hat{K}_c$, $\hat{\tau}_I$ and $\hat{\tau}_D$ depend only on the performance parameters β and ζ.

In order to examine the stability margins of the integrating plus delay process under feedback control with a PID controller, the actual open-loop transfer function is formulated as follows

$$\begin{aligned} G_{ol}(s) &= K_c(1 + \frac{1}{\tau_I s} + \tau_D s)\frac{Ke^{-ds}}{s} \\ &= \hat{K}_c(1 + \frac{1}{\hat{\tau}_I \hat{s}} + \hat{\tau}_D \hat{s})\frac{e^{-\hat{s}}}{\hat{s}} \end{aligned} \tag{7.46}$$

Since the open-loop transfer function is only a function of the performance parameters, the gain and phase margins (GM and PM) of the designed closed-loop system are independent of the actual process parameters. However, the actual critical frequencies are obtained by dividing the normalized critical frequencies with the process delay. We are also able to obtain the relative delay margin as

$$RDM = \frac{\Delta d}{d} = \frac{PM}{\hat{w}_p} \tag{7.47}$$

where $\hat{w}_p$ is the critical phase margin frequency. Note that the right-hand side of Equation (7.47) depends only on the performance parameters β and ζ.

Presentation of the Tuning Rules

From Equations (7.43)-(7.45), the normalized PID controller parameters can be calculated for a given set of performance parameters. In this case, tuning rules are more straightforward to derive than for the first order plus delay case because the normalized $\hat{Y}(j\hat{w})$ is not a function of the process parameters.

In the derivation of the rules, the damping factor ζ has been chosen equal to either 0.707 or 1 to produce two sets of tuning rules, with the parameter β being used to adjust the closed-loop response speed. The desired closed-loop time constant is $\tau = \beta d$. β has been varied for both cases from 1 to 17 and the corresponding normalized PID controller parameters $\hat{K}_c$, $\hat{\tau}_I$ and $\hat{\tau}_D$ have been calculated. From this information, polynomial functions have been fit to produce explicit solutions for the normalized controller parameters as a function of β.

$\zeta = 0.707$:

$$\hat{K}_c = \frac{1}{0.7138\beta + 0.3904} \tag{7.48}$$

$$\hat{\tau}_I = 1.4020\beta + 1.2076 \tag{7.49}$$

$$\hat{\tau}_D = \frac{1}{1.4167\beta + 1.6999} \tag{7.50}$$

$\zeta = 1$:

$$\hat{K}_c = \frac{1}{0.5080\beta + 0.6208} \tag{7.51}$$

$$\hat{\tau}_I = 1.9885\beta + 1.2235 \tag{7.52}$$

$$\hat{\tau}_D = \frac{1}{1.0043\beta + 1.8194} \tag{7.53}$$

The actual controller parameters are obtained using the scaling Equations (7.43)-(7.45), i.e. $K_c = \frac{\hat{K}_c}{dK}$, $\tau_I = d\hat{\tau}_I$ and $\tau_D = d\hat{\tau}_D$. The gain, phase and relative delay margins for these PID tuning rules are shown in Figures 7.14 and 7.16. In general, the choice of damping factor $\zeta = 0.707$ produces smaller controller parameters in comparison to $\zeta = 1$. The former choice also leads to larger gain and relative delay margins, but smaller phase margins, implying a faster closed-loop response.

The settings for PI control with these rules are obtained from the same set of equations for PID with $\tau_D = 0$. The gain, phase and relative delay margins for PI control are shown in Figures 7.15 and 7.17.

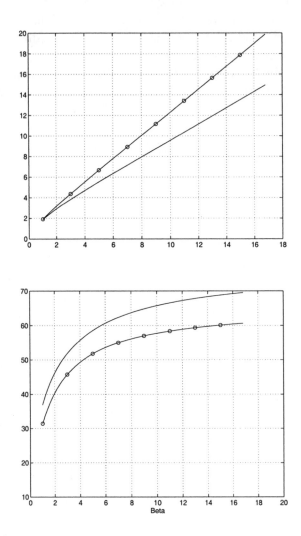

Figure 7.14: *Stability margins for PID tuning rules (solid: $\zeta = 1$; solid with 'o': $\zeta = 0.707$). Upper diagram: gain margins; lower diagram: phase margins*

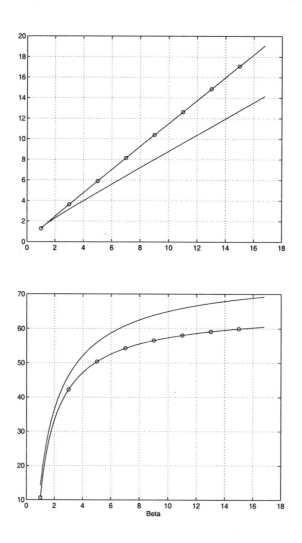

Figure 7.15: *Stability margins for PI tuning rules (solid: $\zeta = 1$; solid with 'o': $\zeta = 0.707$). Upper diagram: gain margins; lower diagram: phase margins*

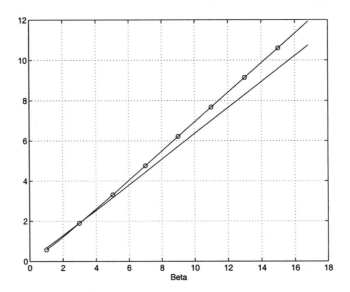

Figure 7.16: *Relative delay margins for PID tuning rules (solid: $\zeta = 1$; solid with 'o': $\zeta = 0.707$)*

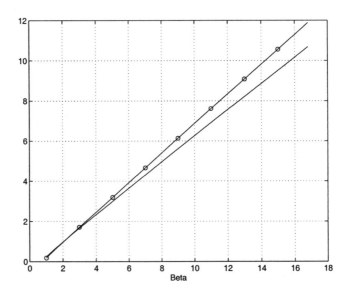

Figure 7.17: *Relative delay margins for PI tuning rules (solid: $\zeta = 1$; solid with 'o': $\zeta = 0.707$)*

Evaluation of the Tuning Rules

Consider the integrating plus delay process used in Tyreus and Luyben (1992) described by the transfer function

$$G(s) = \frac{K}{s} e^{-ds} \tag{7.54}$$

where $K = 0.0506$ and $d = 6$. The desired closed-loop time constant τ is chosen to be equal to d, $2d$, $3d$ and $4d$ corresponding to $\beta = 1$, 2, 3 and 4, and the damping factor is selected to be 0.707. For the first two choices of β, a PID controller is used and for the latter two choices, a PI controller is used. The normalized controller parameters are calculated from Equations (7.48)-(7.50) and then the true parameters are obtained using the scaling Equations (7.43)-(7.45). Figure 7.18 shows the process output and control signal responses for a unit step setpoint change followed by a unit step load disturbance.

In Tyreus and Luyben's (1992) work, their objective was to design a PI controller for integrating plus delay processes. Their performance specification was given in the frequency domain, where the peak value of the magnitude of the closed-loop transfer function $G_{r \to y}$ was chosen to be $+2dB$. A numerical procedure was used to find the normalized PI controller parameters to achieve this performance specification. It is interesting to note that their final tuning rules for the PI parameters are presented in the same fashion as Equations (7.43) and (7.44), except that they are given for only the single performance specification. Their normalized proportional gain $\hat{K}_c$ is equal to 0.487 and their normalized integral time constant $\hat{\tau}_I$ is equal to 8.75.

With our tuning rules, if we choose $\zeta = 1$ and $\beta = 3$, then $\hat{K}_c = 0.466$ and $\hat{\tau}_I = 7.19$. Therefore, for this particular choice of closed-loop performance, our tuning rules correspond almost exactly to those presented by Tyreus and Luyben (1992). The key benefit with our tuning rules is that different values for β, depending on the control objective and desired stability margins, may be selected and used to easily calculate the corresponding PID controller parameters.

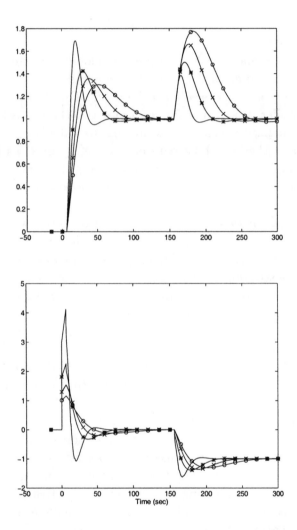

Figure 7.18: *Setpoint and load disturbance responses for Tyreus and Luyben example (solid: $\beta = 1$; solid with '*': $\beta = 2$; solid with '×': $\beta = 3$; solid with 'o': $\beta = 4$). Upper diagram: process output; lower diagram: control signal*

Chapter 8

Recursive Estimation from Relay Feedback Experiments

8.1 INTRODUCTION

This chapter describes two new methods for obtaining frequency response and step response models from processes operating under relay feedback control. Both methods are based on the frequency sampling filter model structure and a recursive least squares estimator.

This chapter contains three sections. Section 8.2 describes the approach to frequency response estimation using data generated from a standard relay experiment. In Section 8.3, a modified relay experiment is proposed for step response estimation.

Portions of this chapter have been reprinted from *Automatica* **35**, L. Wang, M.L. Desarmo and W.R. Cluett, "Real-time estimation of process frequency response and step response from relay feedback experiments", pp. 1427-1436, 1999, with permission from Elsevier Science.

8.2 RECURSIVE FREQUENCY RESPONSE ESTIMATION

Figure 8.1 is a block diagram of a relay feedback control system, where u is the relay output signal and e is the feedback error signal entering the relay element. A simple relay is a nonlinear element that switches between two levels, $-d$ and $+d$, based on the sign of the error signal e and generates a square wave relay output signal u to the process. If the process output y is corrupted with noise, a hysteresis of width ε is added to the relay. Adding

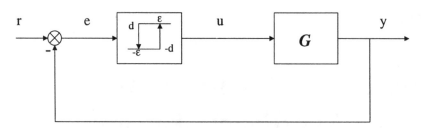

Figure 8.1: *Block diagram for relay feedback control system*

hysteresis to the relay produces a dead-zone which is used to prevent the relay output signal from switching due to the noise. Only when the absolute value of the feedback error signal exceeds ε will the relay output u change sign from its previous value. This nonlinear closed-loop system will exhibit a stable limit cycle for many processes. The period of this limit cycle is determined by the process dynamics, the amplitude of the relay d, and the width of the hysteresis ε. It is well known that if the width of the hysteresis is equal to zero, then the oscillation frequency corresponds approximately to the critical frequency of the process, i.e. the point of intersection of the process Nyquist curve with the negative real axis. As the hysteresis width increases, the oscillation frequency decreases. Other frequencies can be obtained by adding a dynamic element, such as a linear filter, in series with the relay. If the linear dynamic element is a pure integrator, the frequency of oscillation corresponds to the point of intersection of the process Nyquist curve with the negative imaginary axis.

Suppose that the process to be identified is placed under relay feedback control and oscillates with some period T. Using a sampling interval of Δt, the number of samples within a period is $N' = \frac{T}{\Delta t}$. The periodic square wave $u(k)$ generated by the relay output can be completely described over this period $[0, T]$ using a discrete Fourier expansion (Godfrey, 1993)

$$u(k) = \sum_{l=-\frac{N'-1}{2}}^{\frac{N'-1}{2}} A_l e^{j\frac{2\pi lk}{N'}} \tag{8.1}$$

for $k = 0, 1, \ldots, N' - 1$, where

$$A_l = \frac{1}{N'} \sum_{k=0}^{N'-1} u(k) e^{-j\frac{2\pi lk}{N'}} \tag{8.2}$$

for $l = 0, \pm 1, \ldots, \pm \frac{N'-1}{2}$. If the input signal is also symmetric and the time origin is taken at one of the relay switches, $A_l = 0$ for $l = 0, \pm 2, \pm 4, \ldots$. The magnitudes of the nonzero values of A_l decrease with increasing $|l|$.

For frequency response estimation, we set the parameter N in the FSF model in Equation (4.22) to be equal to N' to capture the dominant periodic frequency components in the input and output signals. Thus the process output $y(k)$ can be described by

$$y(k) = \sum_{l=-\frac{n-1}{2}}^{\frac{n-1}{2}} G(e^{j\frac{2\pi l}{N}}) \frac{1}{N} \frac{1 - z^{-N}}{1 - e^{j\frac{2\pi l}{N}} z^{-1}} u(k) + \xi(k) \tag{8.3}$$

where $\xi(k)$ is the disturbance term. We now define the parameter vector to be estimated as

$$\theta = [G(e^{j0}) \ \ G(e^{j\frac{2\pi}{N}}) \ \ G(e^{-j\frac{2\pi}{N}}) \ \ \cdots \ \ G(e^{j\frac{(n-1)\pi}{N}}) \ \ G(e^{-j\frac{(n-1)\pi}{N}})]^T$$

and its corresponding regressor vector as

$$\phi(k) = [f(k)^0 \ \ f(k)^1 \ \ f(k)^{-1} \ \ \cdots \ \ f(k)^{\frac{n-1}{2}} \ \ f(k)^{-\frac{n-1}{2}}]^T$$

where

$$f(k)^r = \frac{1}{N} \frac{1 - z^{-N}}{1 - e^{j\frac{2\pi r}{N}} z^{-1}} u(k) \tag{8.4}$$

for $r = 0, \pm 1, \ldots, \pm \frac{n-1}{2}$. Therefore, the total number of frequencies included in this process output description is $\frac{n-1}{2} + 1$.

If $u(k)$ is a periodic and symmetric signal, the filter outputs for even values of r ($r = 0, \pm 2, \pm 4, \ldots$) are equal to zero after one complete period N. In addition, the magnitudes of the nonzero filter outputs corresponding to $r = \pm 1, \pm 3, \ldots$ decrease as $|r|$ increases. Therefore, in this situation, the only terms required to accurately describe the process output $y(k)$ in Equation (8.3) for processes with a monotonically decreasing frequency response may be those with $r = \pm 1, \pm 3$ and ± 5. However, because output disturbances and measurement noise are encountered in most practical situations, $u(k)$ is seldom an ideal periodic and symmetric signal. In many cases though, $u(k)$ would be nearly periodic and the parameter N could be chosen based on an estimate of the average period. In order to avoid biased estimates of the frequency response parameters in such circumstances, we suggest that all of the terms corresponding to $r = 0, \pm 1, \ldots, \pm 5$ be included in the process description.

Given the process input-output data generated from the relay experiment, the parameter vector θ can be estimated using a recursive algorithm. Here, we propose to use the recursive least squares algorithm (Goodwin and Sin, 1984) given as follows

$$\hat{\theta}(k) = \hat{\theta}(k-1) + P(k-1)\phi(k)(y(k) - \phi(k)^T \hat{\theta}(k-1)) \qquad (8.5)$$

and

$$P(k-1) = P(k-2) - \frac{P^*(k-2)\phi(k)\phi^*(k)P(k-2)}{1 + \phi^*(k)P(k-2)\phi(k)} \qquad (8.6)$$

where (*) denotes the complex conjugate transpose. We also suggest using the process input-output data from the first complete period with a standard batch least squares estimator to initialize $\hat{\theta}(0)$ and $P(-1)$.

Bitmead and Anderson (1981) proposed the use of a recursive least mean squares (LMS) algorithm to estimate frequency response coefficients using the FSF model structure. These authors treated the problem as a collection of independent, or decoupled, estimation problems by assuming that the outputs of the frequency sampling filters satisfy an approximate orthogonality property. This assumption allowed the authors to reduce the larger single estimation problem (Nth-order) to a collection of several smaller (one or two parameter) estimation problems, with their objective being to avoid the dimensionality and ill-conditioning problems known to be associated with the larger problem.

Our choice of the more rapidly converging recursive least squares approach to solve the larger single estimation problem is justified by the fact that we only need to estimate n parameters. Because we can choose $n \ll N$, the dimensionality and ill-conditioning problems can be avoided. Also, the FSF outputs are only truly orthogonal if the input is either white noise or periodic (Goberdhansingh *et al.*, 1992). Therefore, in practice, it is desirable to solve the larger single estimation problem to avoid biased parameter estimates.

One other point worth mentioning is that, if the input is periodic and the data length is an integer multiple of the fundamental period, the frequency estimates obtained from Equations (8.5) and (8.6) would be equivalent to those obtained using a DFT analysis.

Example 8.1. The following transfer function is representative of the dynamics typically associated with paper machine basis weight (EnTech, 1993)

$$G(s) = \frac{e^{-100s}}{45s + 1} \qquad (8.7)$$

The objective is to recursively estimate the process frequency response at the dominant harmonic frequencies generated under a standard relay experiment. A white noise disturbance sequence with unit variance has been added to the output. The relay amplitude was set equal to 2 and the hysteresis level was set equal to 3 (3 times the standard deviation of the output noise). The process was sampled with a time interval of 0.67 seconds. Note that, although the hysteresis level is larger than the noise-free process output response to an input change of magnitude 2, a limit cycle still occurs due to the presence of the noise.

Figure 8.2 shows the process input-output data generated under a standard relay experiment. An estimate of $\frac{N'}{2}$ was taken as the number of samples between the second and third switches of the relay. Figure 8.3 illustrates how the real and imaginary parts of the parameter estimates behave for the first harmonic ($r = 1$) and the third harmonic ($r = 3$) over the duration of the relay experiment. The estimates of these parameters up until the third switch in the relay output (≈ 300 sec) are constant and equal to $\hat{\theta}(0)$. Figure 8.4 compares the two estimated frequency responses after 500 sec with the true process frequency response. These latter two figures show that the parameter estimates have effectively converged after 500 sec. This compares very favourably with the total process settling time of approximately 325 sec.

The example used here is a first order plus delay process with very little roll-off at higher frequencies. For processes with more roll-off at higher frequencies, it may not be realistic to expect accurate identification at frequencies beyond the first harmonic, depending on the signal-to-noise ratio at the higher harmonics.

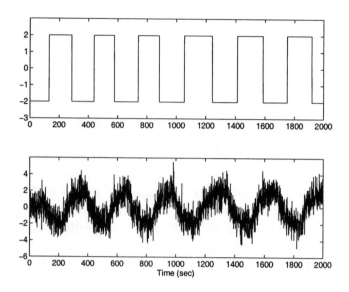

Figure 8.2: *Data generated under a standard relay experiment for Example 8.1.
Upper diagram: relay output; lower diagram: process output*

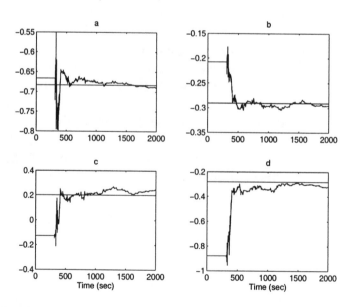

Figure 8.3: *Real and imaginary parts of the frequency response estimates corresponding to the first (a,b) and third (c,d) harmonics, respectively, for Example 8.1*

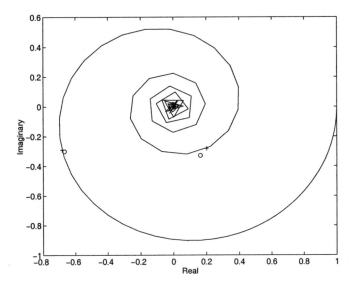

Figure 8.4: *Comparison of true and estimated frequency responses after 500 sec for Example 8.1 (solid with '+': true response; 'o': estimated values)*

8.3   RECURSIVE STEP RESPONSE ESTIMATION

As discussed in the previous section, a standard relay experiment typically produces a limit cycle that is dominated by a single frequency. However, this information is not sufficient for the estimation of an accurate process step response model. This raises the issue of how to generate the appropriate information. One approach is to inject a dither signal while the process is under some sort of feedback control, either additively to the controller output or via the setpoint. However, this requires design of the dither signal, i.e. decisions must be made concerning its power spectrum. Another alternative is to make use of multiple relay experiments to generate frequency response information at several frequencies.

Our objective is to develop a single relay experiment to automatically generate the desired information. The proposed apparatus combines in parallel a relay element and an integrator in series with a relay element. Figure 8.5 provides a block diagram of this apparatus. The experiment is performed by alternatively switching the error signal between the relay path and the integrator-relay path. The design of the experiment then reduces to the selection of this switching sequence. The input signal generated from this type of relay experiment will no longer be dominated by a single frequency but

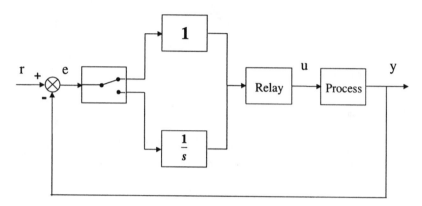

Figure 8.5: *Block diagram for proposed relay apparatus*

will instead contain frequencies over a range corresponding to the process phase shift of $-\frac{\pi}{2}$ and $-\pi$ in the noise-free case, and over a slightly lower frequency range when hysteresis is added to the relay elements.

The input-output data generated from this relay experiment can be used along with Equations (8.5) and (8.6) to recursively estimate the parameters in the FSF model, and then combined with Equation (5.6) to determine an estimate of the process step response model. However, in this case, N in Equation (8.3) must be chosen according to the process settling time. If prior information on the approximate settling time is available, then this may be used to preselect a value for N. However, it would be desirable to find a simple way to estimate even this parameter on-line during the relay experiment. When an integrator is placed in series with a relay that has zero hysteresis, the closed-loop system will oscillate at a frequency corresponding to the point of intersection of the process Nyquist curve with the negative imaginary axis. From a large number of simulation studies, we have observed that, for many processes, the discrete frequency response at $w = \frac{2\pi}{N}$ radians, corresponding to $w = \frac{2\pi}{T_s}$ radians/sec, is located in the vicinity of this point of intersection. Therefore, if the relay experiment begins with the error following the integrator-relay path for at least four successive switches in the relay output such that a steady state oscillation is reached, an estimate of $\frac{N}{2}$ can be taken as the number of samples between the third and fourth switches of the relay. In addition, $\hat{\theta}(0)$ and $P(-1)$ can be estimated at this point time from the previous N sets of input-output data.

The only remaining parameter to be chosen is n, the number of FSF model parameters to be estimated in Equation (8.3). In Chapter 5, a batch generalized least squares algorithm was presented in which the *PRESS* statistic introduced in Chapter 3 was used for selecting n. For the recursive approach proposed here, we suggest fixing the number of parameters to be estimated by preselecting $n = 11$, which we have found to be sufficient for a wide range of processes.

8.3.1 Simulation case study

Here, we study the performance of the proposed methodology for recursive estimation of the process step response using five examples, four of which have been selected from the test batch presented in Åström and Hägglund (1995).

Process A:
$$G(s) = \frac{e^{-s}}{(s+1)^2} \qquad (8.8)$$

Process B:
$$G(s) = \frac{1}{(s+1)^8} \qquad (8.9)$$

Process C:
$$G(s) = \frac{1}{(1+s)(1+0.5s)(1+0.5^2 s)(1+0.5^3 s)} \qquad (8.10)$$

Process D:
$$G(s) = \frac{1 - 0.2s}{(s+1)^3} \qquad (8.11)$$

Process E:
$$G(s) = \frac{e^{-s}}{s^2 + 2 \times 0.45s + 1} \qquad (8.12)$$

Process E is not found in Åström and Hägglund's test batch but we have added it to provide a case with underdamped dynamics.

The quality of the estimated model is measured here by the amount of departure from the true impulse and step response coefficients. More precisely, we use the sum of the squared deviations from the true impulse response coefficients ($E_{impulse}$) and the sum of squared deviations from the true step response coefficients (E_{step}) to quantify the closeness of the fit to the true model, i.e.

$$E_{impulse} = \sum_{i=0}^{N-1} (h_i - \hat{h}_i)^2 \tag{8.13}$$

$$E_{step} = \sum_{i=0}^{N-1} (g_i - \hat{g}_i)^2 \tag{8.14}$$

The impulse response error, $E_{impulse}$, focuses more on the closeness of the estimated process dynamics to the true process dynamics with equal weighting at all frequencies, while the step response error, E_{step}, measures the accuracy of the estimated model with more emphasis in the low frequency range (Dayal and MacGregor, 1996).

For the simulations, a white noise disturbance sequence with variance equal to 0.8^2 has been added to the process output. The relay amplitude has been set equal to 2 and the hysteresis level has been initially set equal to 0.08. The relay experiment is started with the error following the integrator-relay path. Because of the averaging effect of integration, a small hysteresis level can be tolerated and is used here initially to obtain an estimate of the frequency response corresponding to the point of intersection of the process Nyquist curve with the negative imaginary axis, and in turn an estimate of N. The experiment is allowed to proceed until four switches in the relay output have occurred at which time the value of N is estimated. The hysteresis level is then fixed to a value of 2.5 (approximately 3 times the standard deviation of the output noise) and the error is switched to the relay path for one complete period (two switches). Then, the error is switched back-and-forth between the integrator-relay and relay paths, after 2-3 switches in the relay output have occurred along a given path. Total simulation time for each case is approximately four times the process settling time. The sampling rate is adjusted for each case so that 3000 sets of input-output data are collected within the total simulation time.

To illustrate the results, the estimated FSF model parameters have been converted into step response and impulse response coefficients at integer multiples of the process settling time. The values of $E_{impulse}$ and E_{step} are summarized in Table 8.1, along with the estimated T_s values obtained from the relay experiment. These results show that the estimated models are all converging as the length of the relay experiment increases.

Process	After T_s	After $2T_s$	After $3T_s$	After $4T_s$
A ($T_s = 11$ sec)				
$E_{impulse}$	9.11×10^{-4}	1.27×10^{-4}	8.66×10^{-5}	1.04×10^{-4}
E_{step}	26.83	0.66	0.43	0.11
B ($T_s = 31$ sec)				
$E_{impulse}$	8.76×10^{-4}	2.62×10^{-5}	3.45×10^{-5}	3.28×10^{-5}
E_{step}	4.45	3.47	0.68	0.18
C ($T_s = 6.3$ sec)				
$E_{impulse}$	3.57×10^{-4}	8.69×10^{-5}	4.66×10^{-5}	3.25×10^{-5}
E_{step}	2.35	0.69	0.43	0.15
D ($T_s = 19$ sec)				
$E_{impulse}$	1.5×10^{-2}	9.5×10^{-4}	8.49×10^{-4}	8.73×10^{-4}
E_{step}	19.5	0.32	0.24	0.31
E ($T_s = 9.3$ sec)				
$E_{impulse}$	4.31×10^{-4}	8.18×10^{-5}	4.88×10^{-5}	5.08×10^{-5}
E_{step}	0.8	0.32	0.25	0.09

Table 8.1: *Case study results*

Comparison with FIR Model

We have selected Process A from the case study for further study and comparison with results obtained using an FIR model. Figure 8.6 shows the process input-output data collected using the modified relay experiment. The estimated step response models are presented in Figure 8.7 along with the true step response. These plots indicate that as the data length increases the estimated step response models converge to the true step response. In fact, the estimated step response model changes very little after only two settling times of data. Figure 8.8 confirms that the estimated FSF model parameters after two settling times of data are very accurate in the low and medium frequency regions ($r = 0, \pm1, \pm2$) and the deviations in the higher frequency region ($r = \pm3, \pm4, \pm5$) are modest but do not have a significant effect on the step response model accuracy.

For the FIR model estimate, we decided to work with exactly the same set of data collected from the relay experiment. As a result, 662 parameters needed to be estimated based on the estimated value for T_s and the sampling rate for this process (i.e. $N = 662$). A batch least squares algorithm has been used to estimate these FIR model parameters. The estimated step response models are compared with the true step response in Figure 8.9. After

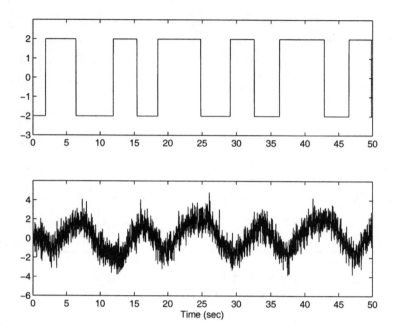

Figure 8.6: *Data generated under modified relay experiment for Process A. Upper diagram: relay output; lower diagram: process output*

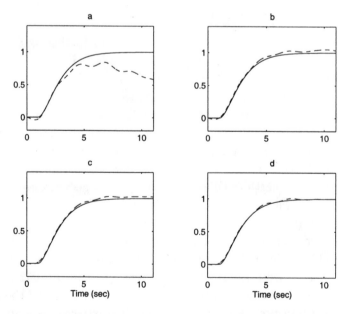

Figure 8.7: *Step response for Process A (solid: true response; dashed: estimated response using FSF model after (a) T_s, (b) $2T_s$, (c) $3T_s$, and (d) $4T_s$)*

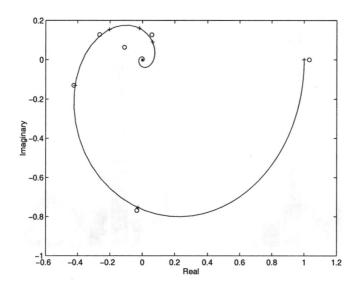

Figure 8.8: *Frequency response for Process A (solid line: continuous-time frequency response; '+': true FSF parameters; 'o': estimated FSF parameters after $2T_s$)*

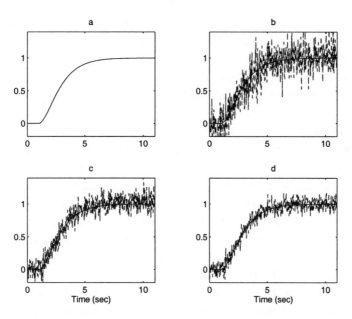

Figure 8.9: *Step response for Process A (solid: true response; dashed: estimated response using FIR model after (a) T_s (no model available), (b) $2T_s$, (c) $3T_s$, and (d) $4T_s$)*

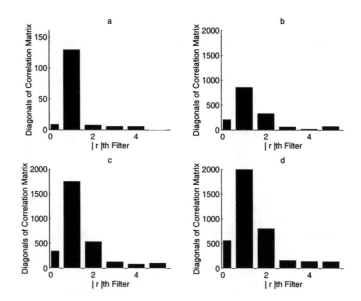

Figure 8.10: *Diagonal elements of the correlation matrix for Process A after (a) T_s, (b) $2T_s$, (c) $3T_s$, and (d) $4T_s$*

one settling time of data, the correlation matrix was singular and therefore we were unable to obtain an estimated model. It is evident from these plots that the estimated step responses obtained from the FIR model are unbiased but are also very noisy and could not be used directly for control system design. It is important to point out that a slower sampling rate would not have improved these results.

This comparison confirms the results presented in Chapter 5, where the use of a reduced order FSF model has been interpreted as a means for obtaining a smooth step response estimate. These results illustrate how the FSF approach improves the numerical conditioning of the parameter estimation problem by neglecting the higher frequency parameters that contribute significantly to the variance error but little to the bias error. It is also worth mentioning again that, with the FSF approach, the diagonal elements of the correlation matrix are proportional to the periodogram of the input signal in the vicinity of the FSF frequencies, and to the data length. Therefore, the accuracy of a particular FSF parameter is directly related to the magnitude of the corresponding diagonal element in the correlation matrix according to Equation (5.25). To examine the energy distribution of the input signal generated by the modified relay experiment for Process A, Figure 8.10 shows the magnitude plots of the diagonal elements of the correlation matrix corresponding to the FSF model with $n = 11$. After one settling time of data, the energy of the input signal is focused only at the first pair of frequencies ($r = \pm 1$). However, after two settling times, energy is clearly present at both the zero frequency ($r = 0$) and at the second pair of frequencies ($r = \pm 2$), and to a lesser degree at the higher frequencies ($r = \pm 3, \pm 4, \pm 5$). This illustrates how the data generated from the proposed experiment leads to accurate estimates of the frequency response in the low and medium frequencies, including the steady state gain and, in turn, an accurate estimate of the step response via the FSF model.

8.3.2 Automated design of an identification experiment

The proposed relay device and FSF algorithm could readily be bundled together as a stand-alone apparatus for on-line process frequency/step response identification and subsequent controller design and tuning (see, for example, Hägglund and Åström, 1985). However, the modified relay experiment on its own provides some interesting and new ideas about how to design input signals for process identification. One of the main benefits of the proposed methodology is that the design of an identification experiment suitable for obtaining an accurate step response model for stable processes has now been automated. A standard approach to the design of an identification experiment involves first the off-line design of the input signal followed by the identification experiment itself. This off-line design typically requires prior information on the dominant process dynamics and/or the desired power spectrum of the input signal. Under the modified relay experiment, the relevant frequency response information is automatically generated without the requirement of any prior process information except the sign of the gain. Here, a general input signal design algorithm suitable for process identification is proposed which does not require an actual relay device for implementation.

Step 1: With the process initially at or near steady state, estimate the process output noise level by calculating its standard deviation (σ) and set the hysteresis width $\varepsilon = 3\sigma$. Preselect the amplitude (d) of the input signal where the input signal will switch $\pm d$ around a nominal value $\bar{u}$. Designate the corresponding nominal value of the process output as the reference output value r.

Step 2: Set the input signal $u(k) = \bar{u} + d$ and start calculating the integrated error signal according to $e_I(k) = e(k)\Delta t + e_I(k-1)$, with $e(k) = r - y(k)$ and the initial value of $e_I = 0$.

Step 3: At each sampling instant, calculate the currrent value of the input signal $u(k)$ according to:
If

$$|e_I(k)| \geq \varepsilon \tag{8.15}$$

then (for positive process gain)

$$u(k) = \bar{u} + d \times sign(e_I(k)) \tag{8.16}$$

then (for negative process gain)

$$u(k) = \bar{u} - d \times sign(e_I(k)) \tag{8.17}$$

else

$$u(k) = u(k-1) \tag{8.18}$$

Step 4: Repeat Step 3 until three switches in the process input have occurred.

Step 5: After the final switch based on Step 3, calculate the current value of the input signal $u(k)$ according to:
If

$$|e(k)| \geq \varepsilon \tag{8.19}$$

then (for positive process gain)

$$u(k) = \bar{u} + d \times sign(e(k)) \tag{8.20}$$

then (for negative process gain)

$$u(k) = \bar{u} - d \times sign(e(k)) \tag{8.21}$$

else

$$u(k) = u(k-1) \tag{8.22}$$

Step 6: Repeat Step 5 until two switches in the process input have occurred.

Step 7: Alternate between Steps 3 and 5 after 2-3 switches in the process input have occurred within a given step. At the beginning of Step 3, always set the initial value of $e_I = 0$.

Bibliography

Åström, K. J. (1991), 'Assessment of achievable performance of simple feedback loops', *Int. J. Adaptive Control and Signal Processing* **5**, 3–19.

Åström, K. J., C. C. Hang, P. Persson & W. K. Ho (1992), 'Towards intelligent PID control', *Automatica* **28**, 1–9.

Åström, K. J., P. Hagander & J. Sternby (1984), 'Zeros of sampled systems', *Automatica* **20**, 31–38.

Åström, K. J. & T. Hägglund (1984), 'Automatic tuning of simple regulators with specifications on phase and amplitude margins', *Automatica* **20**, 645–651.

Åström, K. J. & T. Hägglund (1988), *Automatic Tuning of PID Controllers*, Instrument Society of America, Research Triangle Park, NC.

Åström, K. J. & T. Hägglund (1995), *PID Controllers: Theory, Design, and Tuning*, Instrument Society of America, Research Triangle Park, NC.

Atkinson, K. E. (1989), *An Introduction to Numerical Analysis*, John Wiley and Sons, New York.

Belsley, D. A. (1991), *Conditioning Diagnostics: Collinearity and Weak Data in Regression*, John Wiley and Sons, New York.

Bitmead, R. R. & B. D. O. Anderson (1981), 'Adaptive frequency sampling filters', *IEEE Trans. on Circuits and Systems* **28**, 524–534.

Box, G. E. P. & G. M. Jenkins (1976), *Time Series Analysis: Forecasting and Control*, Holden-Day, San Francisco.

Chien, D. C. H. & A. Penlidis (1994a), 'Effect of impurities on continuous solution methyl methacrylate polymerization reactors. I. Open-loop process identification results', *Polymer Reaction Engineering* **2**, 163–213.

Chien, D. C. H. & A. Penlidis (1994b), 'Effect of impurities on continuous solution methyl methacrylate polymerization reactors-II. Closed-loop real-time control', *Chemical Engineering Science* **49**, 1855–1868.

Chien, I.-L. & P. S. Fruehauf (1990), 'Consider IMC tuning to improve controller performance', *Chem. Eng. Progress* **86**(10), 33–41.

Clowes, G. J. (1965), 'Choice of the time-scaling factor for linear system approximations using orthonormal Laguerre functions', *IEEE Transactions on Automatic Control* **10**, 487–489.

Co, T. B. & B. E. Ydstie (1990), 'System identification using modulating functions and fast Fourier transforms', *Computers and Chemical Engineering* **14**, 1051–1066.

Cohen, G. H. & G. A. Coon (1953), 'Theoretical consideration of retarded control', *Transactions of the A.S.M.E.* **75**, 827–834.

Cutler, C. R. & F. H. Yocum (1991), 'Experience with the DMC inverse for identification', *Proc. Fourth International Conference on Chemical Process Control, Padre Island, Texas*, 297–317.

Dayal, B. S. & J. F. MacGregor (1996), 'Identification of finite impulse response models: methods and robustness issues', *Ind. Eng. Chem. Res.* **35**, 4078–4090.

Desoer, C. A. & M. Vidyasagar (1975), *Feedback Systems: Input-Output Properties*, Academic Press, New York.

EnTech (1993), *Automatic Controller Dynamic Specification*, EnTech Control Engineering Inc., Toronto.

Fertik, H. A. (1975), 'Tuning controllers for noisy processes', *ISA Transactions* **14**, 292–304.

Goberdhansingh, E., L. Wang & W. R. Cluett (1992), 'Robust frequency domain identification', *Chemical Engineering Science* **47**, 1989–1999.

Godfrey, K. (1993), *Perturbation Signals for System Identification*, Prentice Hall, New York.

Goodwin, G. C. & K. S. Sin (1984), *Adaptive Filtering Prediction and Control*, Prentice Hall, Englewood Cliffs, NJ.

Goodwin, G. C., M. Gevers & B. Ninness (1992), 'Quantifying the error in estimated transfer functions with application to model order selection', *IEEE Transactions on Automatic Control* **37**, 913–928.

Goodwin, G. C. & R. L. Payne (1977), *Dynamic System Identification: Experiment Design and Data Analysis*, Academic Press, New York.

Hägglund, T. & K. J. Åström (1985), 'Method and an apparatus in tuning a PID-regulator', *United States Patent* **4549123**.

Hang, C. C., K. J. Åström & W. K. Ho (1991), 'Refinements of the Ziegler-Nichols tuning formula', *IEE Proceedings-D* **138**, 111–118.

Harris, T. J. (1989), 'Assessment of control loop performance', *Can. J. Chem. Eng.* **67**, 856–861.

Harris, T. J. & B. D. Tyreus (1987), 'Comments on "Internal model control. 4. PID controller design"', *Ind. Eng. Chem. Res.* **26**, 2161–2162.

Korenberg, M., S. A. Billings, Y. P. Liu & P. J. McIlroy (1988), 'Orthogonal parameter estimation algorithm for non-linear stochastic systems', *Int. J. Control* **48**, 193–210.

Kosut, R. L. & B. D. O. Anderson (1994), 'Least-squares parameter set estimation for robust control design', *Proc. American Control Conference, Baltimore, MD*, 3002–3006.

Kreyszig, E. (1988), *Advanced Engineering Mathematics*, John Wiley and Sons, New York.

Lee, Y. W. (1960), *Statistical Theory of Communication*, John Wiley and Sons, New York.

Levin, M. J. (1960), 'Optimum estimation of impulse response in the presence of noise', *IRE Trans. Circuit Theory* **7**, 50–56.

Li, W., E. Eskinat & W. L. Luyben (1991), 'An improved autotune identification method', *Ind. Eng. Chem. Res.* **30**, 1530–1541.

Lilja, M. (1990), 'A frequency domain method for low order controller design', *IFAC 11th Triennial World Congress, Tallinn*, 223–228.

Ljung, L. (1987), *System Identification: Theory for the User*, Prentice Hall, Englewood Cliffs, New Jersey.

Luyben, W. L. (1987), 'Derivation of transfer functions for highly nonlinear distillation columns', *Ind. Eng. Chem. Res.* **26**, 2490–2495.

MacGregor, J. F. & D. T. Fogal (1995), 'Closed-loop identification: the role of the noise model and prefilters', *Journal of Process Control* **5**, 163–171.

MacGregor, J.F., T. Kourti & J. V. Kresta (1991), 'Multivariate identification: a study of several methods', *IFAC Advanced Control of Chemical Processes, Toulouse, France*, 101–107.

Mäkilä, P. M. (1990), 'Laguerre series approximation of infinite dimensional systems', *Automatica* **26**, 985–995.

Middleton, R. H. (1988), 'Frequency domain adaptive control', *IFAC Workshop on Robust Adaptive Control, Newcastle, Australia*, 200–205.

Middleton, R. H. & G. C. Goodwin (1990), *Digital Control and Estimation: A Unified Approach*, Prentice Hall, Englewood Cliffs, NJ.

Myers, R. H. (1990), *Classical and Modern Regression with Applications*, PWS-KENT, Boston.

Olivier, P. D. (1992), 'Approximating irrational transfer functions using Lagrange interpolation formula', *IEE Proceedings-D* **139**, 9–12.

Parker, P. J. & R. R. Bitmead (1987), 'Adaptive frequency response identification', *Proc. 26th Conference on Decision and Control, Los Angeles, CA*, 348–353.

Partington, J. R., K. Glover, H. J. Zwart & R. F. Curtain (1988), 'L_∞ approximation and nuclearity of delay systems', *Systems and Control Letters* **10**, 59–65.

Rake, H. (1980), 'Step response and frequency response methods', *Automatica* **16**, 519–526.

Ricker, N. L. (1988), 'The use of biased least-squares estimators for parameters in discrete-time pulse-response models', *Ind. Eng. Chem. Res.* **27**, 343–350.

Rivera, D. E., M. Morari & S. Skogestad (1986), 'Internal model control. 4. PID controller design', *Ind. Eng. Chem. Process Des. Dev.* **25**, 252–265.

Schei, T.S. (1994), 'Automatic tuning of PID controllers based on transfer function estimation', *Automatica* **30**, 1983–1989.

Schroeder, M.R. (1970), 'Synthesis of low-peak-factor signals and binary sequences with low autocorrelation', *IEEE Trans. Information Theory* **16**, 85–89.

Tulleken, H. J. A. F. (1990), 'Generalized binary nosie test-signal concept for improved identification-experiment design', *Automatica* **26**, 37–49.

Tyreus, B. D. & W. L. Luyben (1992), 'Tuning PI controllers for integrator/dead time processes', *Ind. Eng. Chem. Res.* **31**, 2625–2628.

Unbehauen, H. & G. P. Rao (1987), *Identification of Continuous Systems*, North-Holland, Amsterdam.

Wahlberg, B. (1991), 'System identification using Laguerre models', *IEEE Transactions on Automatic Control* **36**, 551–562.

Wahlberg, B. & L. Ljung (1992), 'Hard frequency-domain model error bounds from least-squares like identification techniques', *IEEE Transactions on Automatic Control* **37**, 900–912.

Wylie, C. R. (1960), *Advanced Engineering Mathematics*, McGraw-Hill, New York.

Zervos, C. C. & G. A. Dumont (1988), 'Deterministic adaptive control based on Laguerre series representation', *Int. J. Control* **48**, 2333–2359.

Zhuang, M. & D. P. Atherton (1993), 'Automatic tuning of optimum PID controllers', *IEE Proceedings-D* **140**, 216–224.

Ziegler, J. G. & N. B. Nichols (1942), 'Optimum settings for automatic controllers', *Transactions of the A.S.M.E.* **64**, 759–768.

Index